T0219695

Processing Big Data with Azure HDInsight

Building Real-World Big Data
Systems on Azure HDInsight
Using the Hadoop Ecosystem

Vinit Yadav

Apress®

Processing Big Data with Azure HDInsight

Vinit Yadav
Ahmedabad, Gujarat, India

ISBN-13 (pbk): 978-1-4842-2868-5   ISBN-13 (electronic): 978-1-4842-2869-2
DOI 10.1007/978-1-4842-2869-2

Library of Congress Control Number: 2017943707

Cover image designed by Freepik

Managing Director: Welmoed Spahr
Editorial Director: Todd Green
Acquisitions Editor: Celestin Suresh John
Development Editor: Poonam Jain and Laura Berendson
Technical Reviewer: Dattatrey Sindol
Coordinating Editor: Sanchita Mandal
Copy Editor: Kim Burton-Weisman
Compositor: SPi Global
Indexer: SPi Global
Artist: SPi Global

Distributed to the book trade worldwide by Springer Science+Business Media New York, 233 Spring Street, 6th Floor, New York, NY 10013. Phone 1-800-SPRINGER, fax (201) 348-4505, e-mail orders-ny@springer-sbm.com, or visit www.springeronline.com. Apress Media, LLC is a California LLC and the sole member (owner) is Springer Science + Business Media Finance Inc (SSBM Finance Inc). SSBM Finance Inc is a **Delaware** corporation.

For information on translations, please e-mail rights@apress.com, or visit http://www.apress.com/rights-permissions.

Apress titles may be purchased in bulk for academic, corporate, or promotional use. eBook versions and licenses are also available for most titles. For more information, reference our Print and eBook Bulk Sales web page at http://www.apress.com/bulk-sales.

Any source code or other supplementary material referenced by the author in this book is available to readers on GitHub via the book's product page, located at www.apress.com/978-1-4842-2868-5. For more detailed information, please visit http://www.apress.com/source-code.

Printed on acid-free paper

Contents at a Glance

Contents

About the Author

Vinit Yadav is the founder and CEO of Veloxcore, a company that helps organizations leverage big data and machine learning. He and his team at Veloxcore are actively engaged in developing software solutions for their global customers using agile methodologies. He continues to build and deliver highly scalable big data solutions.

Vinit started working with Azure when it first came out in 2010, and since then, he has been continuously involved in designing solutions around the Microsoft Azure platform.

Vinit is also a machine learning and data science enthusiast, and a passionate programmer. He has more than 12 years of experience in designing and developing enterprise applications using various .NET technologies.

On a side note, he likes to travel, read, and watch sci-fi. He also loves to draw, paint, and create new things. Contact him on Twitter (@vinityad), or by email (vinit@veloxcore.com), or on LinkedIn (www.linkedin.com/in/vinityadav/).

About the Technical Reviewer

Dattatrey Sindol (a.k.a. Datta) is a data enthusiast. He has worked in data warehousing, business intelligence, and data analytics for more than a decade. His primary focus is on Microsoft SQL Server, Microsoft Azure, Microsoft Cortana Intelligence Suite, and Microsoft Power BI. He also works in other technologies within Microsoft's cloud and big data analytics space.

Currently, he is an architect at a leading digital transformation company in India. With his extensive experience in the data and analytics space, he helps customers solve real-world business problems and bring their data to life to gain valuable insights. He has published numerous articles and currently writes about his learnings on his blog at http://dattatreysindol.com. You can follow him on Twitter (@dattatreysindol), connect with him on LinkedIn (https://www.linkedin.com/in/dattatreysindol), or contact him via email (dattasramblings@gmail.com).

Acknowledgments

Many people have contributed to this book directly or indirectly. Without the support, encouragement, and help that I received from various people, it would have not been possible for me to write this book. I would like to take this opportunity to thank those people.

Writing this book was a unique experience in itself and I would like to thank Apress team to support me throughout the writing. I also want to thank Vishal Shukla, Bhavesh Shah, and Pranav Shukla for their suggestions and continued support, not only for the book but also for mentoring and helping me always. I would like to express my gratitude toward my colleagues: Hardik Mehta, Jigar Shah, Hugh Smith, and Jayesh Mehta, who encouraged me to do better.

I would like to specially thank my wife, Anju, for supporting me and pushing me to give my best. Also, a heartfelt thank-you to my family and friends, who shaped me into who I am today. And last but not least, my brother, Bhavani, for the support and encouragement he always gave me to achieve my dreams.

Introduction

Why this Book?

Hadoop has been the base for most of the emerging technologies in today's big data world. It changed the face of distributed processing by using commodity hardware for large data sets. Hadoop and its ecosystem were used in Java, Scala, and Python languages. Developers coming from a .NET background had to learn one of these languages. But not anymore. This book solely focuses on .NET developers and uses C# as the base language. It covers Hadoop and its ecosystem components, such as Pig, Hive, Storm, HBase, and Spark, using C#. After reading this book, you—as a .NET developer—should be able to build end-to-end big data business solutions on the Azure HDInsight platform.

Azure HDInsight is Microsoft's managed Hadoop-as-a-service offering in the cloud. Using HDInsight, you can get a fully configured Hadoop cluster up and running within minutes. The book focuses on the practical aspects of HDInsight and shows you how to use it to tackle real-world big data problems.

Who Is this Book For?

The audience for this book includes anyone who wants to kick-start Azure HDInsight, wants to understand its core fundamentals to modernize their business, or who wants to get more value out of their data. Anyone who wants to have a solid foundational knowledge of Azure HDInsight and the Hadoop ecosystem should take advantage of this book. The focus of the book appeals to the following two groups of readers.

- Software developers who come from a .NET background and want to use big data to build end-to-end business solutions. Software developers who want to leverage Azure HDInsight's managed offerings in building their next big data project.

- Data scientists and data analysts who want to use Azure HDInsight's capabilities to quickly build big data solutions.

What Will You Learn?

All the code samples are focused from the .NET developer perspective. The following topics are covered in detail.

- The fundamentals of HDInsight and Hadoop, along with its ecosystem

- Provisioning an HDInsight cluster for different types of workloads

- Getting data in/out of an HDInsight cluster and running a MapReduce job on it

- Using Apache Pig and Apache Hive to query data stored inside HDInsight

- Working with HBase, a NoSQL database

- Using Apache Storm to carry out real-time stream analysis

- Working with Apache Spark for interactive, batch, and stream processing

How this Book Is Organized

This book has eight chapters. The following is a sneak peek of the chapters.

Chapter 1: This chapter covers the basics of big data, its history, and explains Hadoop. It introduces the Azure HDInsight service and the Hadoop ecosystem components available on Azure HDInsight, and explains the benefits of Azure HDInsight over other Hadoop distributions.

Chapter 2: The aim of this chapter is to get readers familiar with Azure's offerings, show how to start an Azure subscription, and learn about the different workloads and types of HDInsight clusters.

Chapter 3: This chapter covers Azure blob storage, which is the default storage layer for HDInsight. After that, chapter looks at the different ways to work with HDInsight to submit MapReduce jobs. Finally, it covers Avro library integration.

Chapter 4: The focus of this chapter is to provide understanding of Apache Hive. First, the chapter covers Hive fundamentals, and then dives into working with Hive on HDInsight. It also describes how data scientists using HDInsight can connect with a Hive data store from popular dashboard tools like Power BI or ODBC-based tools. And finally, it covers writing user-defined functions in C#.

Chapter 5: Apache Pig is a platform to analyze large data sets using the procedural language known as Pig Latin, which is covered in this chapter. You learn to use Pig in HDInsight.

Chapter 6: This chapter covers Apache HBase, a NoSQL database on top of Hadoop. This chapter looks into the HBase architecture, HBase commands, and reading and writing data from/to HBase tables using C# code.

Chapter 7: Real-time stream analytics are covered in this chapter. Apache Storm in HDInsight is used to build a stream processing pipeline using C#. This chapter also covers Storm's base architecture and explains different components related to Storm, while giving a sound fundamental overview.

Chapter 8: This chapter focuses on Apache Spark. It explores overall Spark architecture, components, and ways to utilize Spark, such as the batch query, interactive query, stream processing, and more. It then dives deeply into code using Python notebooks and building Spark programs to process data with Mobius and C#.

To get the most out of this book, follow along with the sample code and do the hands-on programs directly in Sandbox or an Azure HDInsight environment.

About versions used in this book: Azure HDInsight changes very rapidly and comes in the form of Azure service updates. Also, HDInsight is a Hadoop distribution from Hortonworks; hence, it also introduces a new version when available. The basics covered in this book will be useful in upcoming versions too.

Happy coding.

■ ■ ■

## CHAPTER 1

Big Data, Hadoop, and HDInsight

Azure HDInsight is a managed Hadoop distribution, developed in partnership with Hortonworks and Microsoft. It uses the Hortonworks Data Platform (HDP) Hadoop distribution, which means that HDInsight is entirely Apache Hadoop on Azure. It deploys and provisions managed Apache Hadoop clusters in the cloud on Windows or Linux machines, which is a unique capability. It provides the Hadoop Distributed File System (HDFS) for reliable data storage. It uses the MapReduce programming model to process, analyze, and report on data stored in distributed file systems. Because it is a managed offering, within a few hours an enterprise can be up and running with a fully configured Hadoop cluster and other Hadoop ecosystem components, such as HBase, Apache Spark, and Apache Storm.

This chapter looks at history so that you understand what big data is and the approaches used to handle large data. It also introduces Hadoop and its components, and HDInsight.

What Is Big Data?

Big data is not a buzzword anymore. Enterprises are adopting, building, and implementing big-data solutions. By definition, big data describes any large body of digital information. It can be historical or in real time, and ranges from streams of tweets to customer purchase history, and from server logs to sensor data from industrial equipment. It all falls under *big data*. As far as the definition goes, there are many different interpretations. One that I like comes from Gartner, an information technology research and advisory company: "Big data is high-volume, high-velocity and/or high-variety information assets that demand cost-effective, innovative forms of information processing that enable enhanced insight, decision making, and process automation." (www.gartner.com/it-glossary/big-data/) Another good description is by Forrester: "Big Data is techniques and technologies that make handling of data at extreme scale economical." (http://blogs.forrestor.com)

© Vinit Yadav 2017
V. Yadav, *Processing Big Data with Azure HDInsight*, DOI 10.1007/978-1-4842-2869-2_1

Based on the preceding definitions, the following are the three Vs of big data.

- **Volume**: The amount of data that cannot be stored using scale-up/vertical scaling techniques due to physical and software limitations. It requires a scale-out or a horizontal scaling approach.

- **Variety**: When new data coming in has a different structure and format than what is already stored, or it is completely unstructured or semi-structured, this type of data is considered a *data variety* problem.

- **Velocity**: The rate at which data arrives or changes. When the window of processing data is comparatively small, then it is called a *data velocity* problem.

Normally, if you are dealing with more than one V, you need a big data solution; otherwise, traditional data management and processing tools can do the job very well. With large volumes of structured data, you can use a traditional relational database management system (RDBMS) and divide the data onto multiple RDBMS across different machines—allowing you to query all the data at once. This process is called *sharding*. Variety can be handled by parsing the schema using custom code at the source or destination side. Velocity can be treated using Microsoft SQL Server StreamInsight. Hence, think about your needs before you decide to use a big data solution for your problem.

We are generating data at breakneck speed. The problem is not with the storage of data, as storage costs are at an all-time low. In 1990, storage costs were around $10K for a GB (gigabyte), whereas now it is less than $0.07 per GB. A commercial airplane has so many sensors installed in it that every single flight generates over 5TB (terabyte) of data. Facebook, YouTube, Twitter, and LinkedIn are generating many petabytes worth of data each day.

With the adoption of Internet of Things (IoT), more and more data is being generated, not to mention all the blogs, websites, user click streams, and server logs. They will only add up to more and more data. So what is the issue? The problem is the amount of data that gets analyzed: large amounts of data are not easy to analyze with traditional tools and technology. Hadoop changed all of this and enabled us to analyze massive amounts of data using commodity hardware. In fact, until the cloud arrived, it was not economical for small and medium-sized businesses to purchase all the hardware required by a moderately sized Hadoop cluster. The cloud really enabled everyone to take advantage of on-demand scaling. Now if you want to analyze terabytes of data, you just spin up a cluster, tear it down when done processing, and pay only for the time that you used the hardware. This has really reduced the overall cost of data processing and has made it available to everyone. Now the actual question is this: How do you build a big data solution? Let's look at the approaches taken so far.

The Scale-Up and Scale-Out Approaches

Traditionally, data is stored in a single processing unit and all requests go through this system only. Once this unit reaches its limit in terms of storage, processing power, or memory, a higher-powered system usually replaces it. This process of expanding a system by adding more resources is called *scale-up,* or *vertical scaling.* The same approach has been used for years to tackle performance improvement issues: add more capable hardware—and performance will go up. But this approach can only go so far; at some point, data or query processing will overwhelm the hardware and you have to upgrade the hardware again. As you scale up, hardware costs begin to rise. At some point, it will no longer be cost effective to upgrade.

Think of a hotdog stand, where replacing a slow hotdog maker with a more experienced person who prepares hotdogs in less time, but for higher wages, improves efficiency. Yet, it can be improved up to only certain point, because the worker has to take their time to prepare the hotdogs no matter how long the queue is and he cannot serve the next customer in the queue until current one is served. Also, there is no control over customer behavior: customers can customize their orders, and payment takes each customer a different amount of time. So scaling up can take you so far, but in the end, it will start to bottleneck.

So if your resource is completely occupied, add another person to the job, but not at a higher wage. You should double the performance, thereby linearly scaling the throughput by distributing the work across different resources.

The same approach is taken in large-scale data storage and processing scenarios: you add more commodity hardware to the network to improve performance. But adding hardware to a network is a bit more complicated than adding more workers to a hotdog stand. These new units of hardware should be taken into account. The software has to support dividing processing loads across multiple machines. If you only allow a single system to process all the data, even if it is stored on multiple machines, you will hit the processing power cap eventually. This means that there has to be a way to distribute not only the data to new hardware on the network, but also instructions on how to process that data and get results back. Generally, there is a master node that instructs all the other nodes to do the processing, and then it aggregates the results from each of them. The scale-out approach is very common in real life—from overcrowded hotdog stands to grocery stores queues, everyone uses this approach. So in a way, big data problems and their solutions are not so new.

Apache Hadoop

Apache Hadoop is an open source project, and undoubtedly the most used framework for big data solutions. It is a very flexible, scalable, and fault-tolerant framework that handles massive amounts of data. It is called a *framework* because it is made up of many components and evolves at a rapid pace. Components can work together or separately, if you want them to. Hadoop and its component are discussed in accordance with HDInsight in this book, but all the fundamentals apply to Hadoop in general, too.

A Brief History of Hadoop

In 2003, Google released a paper on scalable distributed file systems for large distributed data-intensive applications. This paper spawned "MapReduce: Simplified Data Processing on Large Clusters" in December 2004. Based on these papers' theory, an open source project started—Apache Nutch. Soon thereafter, a Hadoop subproject was started by Doug Cutting, who worked for Yahoo! at the time. Cutting named the project Hadoop after his son's toy elephant.

The initial code factored out of Nutch consisted of 5,000 lines of code for HDFS and 6,000 lines of code for MapReduce. Since then, Hadoop has evolved rapidly, and at the time of writing, Hadoop v2.7 is available.

The core of Hadoop is HDFS and the MapReduce programming model. Let's take a look at them.

HDFS

The Hadoop Distributed File System is an abstraction over a native file system, which is a layer of Java-based software that handles data storage calls and directs them to one or more data nodes in a network. HDFS provides an application programming interface (API) that locates the relevant node to store or fetch data from.

That is a simple definition of HDFS. It is actually more complicated. You have large file that is divided into smaller chunks—by default, 64 MB each—to distribute among data nodes. It also performs the appropriate replication of these chunks. Replication is required, because when you are running a one-thousand-nodes cluster, any node could have hard-disk failure, or the whole rack could go down; the system should be able to withstand such failures and continue to store and retrieve data without loss. Ideally, you should have three replicas of your data to achieve maximum fault tolerance: two on the same rack and one off the rack. Don't worry about the name node or the data node; they are covered in an upcoming section.

HDFS allows us to store large amounts of data without worrying about its management. So it solves one problem for big data, while it creates another problem. Now, the data is distributed so you have to distribute processing of data as well. This is solved by MapReduce.

MapReduce

MapReduce is also inspired by the Google papers that I mentioned earlier. Basically, MapReduce moves the computing to the data nodes by using the Map and Reduce paradigm. It is a framework for processing parallelizable problems, spanning multiple nodes and large data sets. The advantage of MapReduce is that it processes data where it resides, or nearby; hence, it reduces the distance over which the data needs to be transmitted. MapReduce is twofold process of distributing computing loads. The first one is Map, which finds all the data nodes where it needs to run the compute and moves the work to those nodes, the second phase is reduce in which the system brings the intermediate results back together and computes them. MapReduce engines have many different implementations (this is discussed in upcoming chapters).

To understand how MapReduce works, take a look at Figure 1-1, which presents a distributed word-count problem that is solved using the MapReduce framework. Let's assume that ABC, DCD, and DBC are stored on different nodes Figure 1-1 shows that first, input data is loaded and divided based on key/value pairs on which mapping is performed on individual nodes. The output of this process is intermediate key/value pairs (i.e., List (K2, V2)). Afterward, this list is given to the reducer, and all similar keys are processed at the same reducer (i.e., K2, List (V2)). Finally, all the shuffling output is combined to form a final list of key/value pairs (i.e., List (K3, V3)).

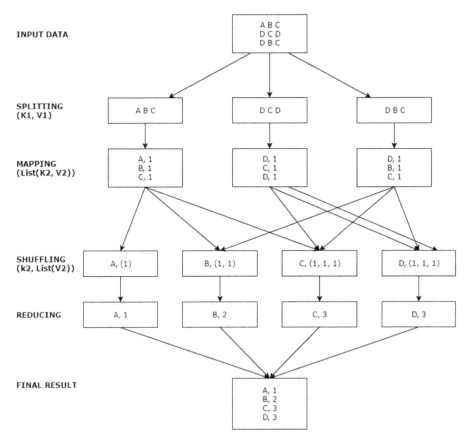

Figure 1-1. *MapReduce word count process*

YARN

YARN stands for *yet another resource negotiator*. It does exactly what it says. YARN acts as a central operating system by providing resource management and application lifecycle management. A central platform to deliver consistent operations, security, and data governance tools across Hadoop clusters. It is a major step in Hadoop 2.0. Hortonworks describes YARN as follows: "YARN is the architectural center of Hadoop that allows multiple data processing engines such as interactive SQL, real-time streaming, data science and batch processing to handle data stored in a single platform, unlocking an entirely new approach to analytics." (`http://hortonworks.com/apache/yarn`). This means that with YARN, you are not bound to use only MapReduce, but you can easily plug current and future engines—for graph processing of a social media website, for example. Also, if you want, you can get custom ISV engines. You can write your own engines as well. In Figure 1-2, you can see all the different engines and applications that can be used with YARN.

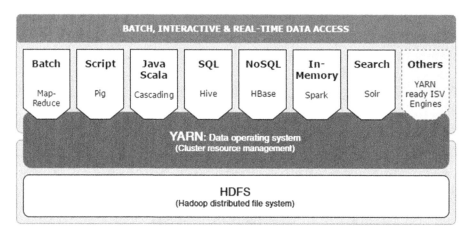

Figure 1-2. YARN applications

Hadoop Cluster Components

Figure 1-3 shows the data flow between a name node, the data nodes, and an HDFS client.

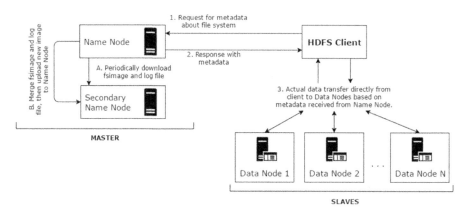

Figure 1-3. Data transfer from client to HDFS

A typical Hadoop cluster consists of following components.

- **Name node**: The head node or master node of a cluster that keeps the metadata. A client application connects to the name node to get metadata information about the file system, and then connects directly to data nodes to transfer data between the client application and data nodes. Here, the name node keeps track of data blocks on different data nodes. The name node is also responsible for identifying dead nodes, decommissioning nodes, and replicating data blocks when needed, like in case of a data node failure. It ensures that the configured replication factor is maintained. It does this through heartbeat signals, which each data node sends to the name node periodically, along with their block reports, which contain data block details. In Hadoop 1.0, the name node is the single point of failure; whereas in Hadoop, 2.0 there is also a secondary name node.

- **Secondary name node**: A secondary name node is in a Hadoop cluster, but its name is bit misleading because it might be interpreted as a backup name node when the name node goes down. The name node keeps track of all the data blocks through a metadata file called fsimage. The name node merges log files to fsimage when it starts. But the name node doesn't update it after every modification of a data block; instead, another log file is maintained for each change. The secondary name node periodically connects to the name node and downloads log files as well as fsimage, updates the fsimage file, and writes it back to the name node. This frees the name node from doing this work, allowing the name node to restart very quickly; otherwise, in a restart, the name node has to merge all the logs since the last restart, which may take significant time. The secondary name node also takes a backup of the fsimage file from the name node.

  In Hadoop 1.0, the name node is a single point of failure, because if it goes down, then there will be no HDFS location to read data from. Manual intervention is required to restart the process or run a separate machine, which takes time. Hadoop 2.0 addresses this issue with the HDFS high-availability feature, where you get the option to run another name node in an active-passive configuration with a hot standby. In an active name node failure situation, the standby takes over and it continues to service requests from the client application.

- **Data node**: A data node stores actual HDFS data blocks. It also stores replicas of data blocks to provide fault tolerance and high availability.

- **JobTracker and TaskTracker**: JobTracker processes
  MapReduce jobs. Similar to HDFS, MapReduce has a master/
  slave architecture. Here, the master is JobTracker and the slave
  is TaskTracker. JobTracker pushes out the work to TaskTracker
  nodes in a cluster, keeping work as close to the data as possible.
  To do so, it utilizes rack awareness: if work cannot be started on
  the actual node where the data is stored, then priority is given
  to a nearby or the same rack node. JobTracker is responsible
  for the distribution of work among TaskTracker nodes. On the
  other hand, TaskTracker is responsible for instantiating and
  monitoring individual MapReduce work. TaskTracker may fail
  or time out; if this happens, only part of the work is restarted. To
  keep TaskTracker work restartable and separate from the rest of
  the environment, TaskTracker starts a new Java virtual machine
  process to do the job. It is TaskTracker's job to send a status
  update on the assigned chunk of work to JobTracker; this is done
  using heartbeat signals that are sent every few minutes.

Figure 1-4 shows a JobTracker flow of submitting a job to TaskTrackers. Client
applications submit jobs to JobTracker. JobTracker requests metadata about the
data files required to complete the job, and then gets the location of the data nodes.
JobTracker chooses the closest TaskTracker to the data and submits part of the job to it.
TaskTracker continuously sends heartbeats. If there is any issue with the TaskTracker,
and no heartbeat is received after a certain amount time, then JobTracker assumes that
the TaskTracker is down and resubmits the job to a different TaskTracker, keeping data
locality and rack awareness in mind. After all the TaskTrackers have finished their jobs,
they submit their results to JobTracker, which then submits it to the client application.

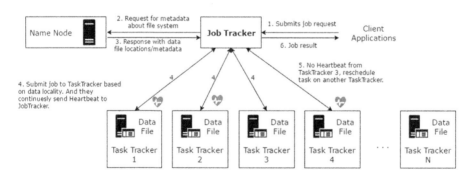

Figure 1-4. *JobTracker and TaskTracker flow*

HDInsight

HDInsight is a Hadoop distribution from Hortonworks. It is a common misconception
that HDInsight is wrapped around Hadoop or is port of core Hadoop by Microsoft,
but it is not. HDInsight is Apache Hadoop running on Azure. It is regular, open source

Hadoop—not a special Microsoft version of Hadoop. Hence, you can run any Hadoop application over HDInsight without modifying it. HDInsight is available on the cloud as a go-to solution for big data analysis. It includes the implementation of Apache products: Spark, HBase, Storm, Pig, Hive, Sqoop, Oozie, Ambari, and so forth. Not only that, HDInsight can be integrated with BI (business intelligence) tools, such as Power BI, Microsoft Excel, SQL Server Analysis Services, and SQL Server Reporting Services.

HDInsight provides a preconfigured Hadoop cluster with other components. At the time of writing, there are a number of options, including Apache Hadoop, Spark, Storm, HBase, Spark, R Server, Interactive Hive (preview), and Kafka (preview). Not all cluster types are available on Windows. Only Hadoop, HBase, and Storm are available on both operating systems, which will change in the future. You can also customize your cluster as you wish. To do this, HDInsight provides *script actions*. Using script actions, you can install components, such as Hue, R, Giraph, Solr, and so forth. These scripts are nothing but bash scripts, and can run during cluster creation on a running cluster, or when adding more nodes to a cluster using dynamic scaling.

Hadoop is generally preferred by Hadoop ecosystem of components, which includes Apache HBase, Apache Spark, Apache Storm, and others. The following are a few of the most useful components under Hadoop umbrella.

- **Ambari**: Apache Ambari is used for provisioning, managing, and monitoring Hadoop clusters. It simplifies the management of a cluster by providing an easy-to-use web UI. Also, it provides a robust API to allow developers to better integrate it in their applications. Note that web UI is only available on Linux clusters; for Windows clusters, REST APIs are the only option.

- **Avro (Microsoft .NET Library for Avro)**: Microsoft Avro Library implements the Apache Avro data serialization system for the Microsoft .NET environment. Avro uses JSON (JavaScript Object Notation) to define a language-agnostic scheme, which means that data serialized in one language can be read by other. Currently, it supports C, C++, C#, Java, PHP, Python, and Ruby. To make a schema available to deserializers, it stores it along with the data in an Avro data container file.

- **Hive**: Most developers and BI folks already know that SQL and Apache Hive were created to enable those with SQL knowledge to submit MapReduce jobs using a SQL-like language called HiveQL. Hive is an abstraction layer over MapReduce. HiveQL queries are internally translated into MapReduce jobs. Hive is conceptually closer to relational databases; hence, it is suitable for structured data. Hive also supports user-defined functions on top of HiveQL for special-purpose processing.

- **HCatalog**: Apache HCatalog is an abstraction layer that presents a relational view of data in the Hadoop cluster. You can have Pig, or Hive, or any other higher-level processing tools on top of HCatalog. It supports the reading or writing of any file for which SerDe (serializer-deserializer) can be written.

- **Oozie**: Apache Oozie is a Java web application that does workflow coordination for Hadoop jobs. In Oozie, a workflow is defined as directed acyclic graphs (DAGs) of actions. It supports different types of Hadoop jobs, such as MapReduce, Streaming, Pig, Hive, Sqoop, and more. Not only these, but also system-specific jobs, such as shell scripts and Java programs.

- **Pig**: Apache Pig is a high-level platform for analyzing large data sets. It requires complex MapReduce transformations that use a scripting language called Pig Latin. Pig translates Pig Latin scripts to a series of MapReduce jobs to run in the Hadoop environment. It automatically optimizes execution of complex tasks, allowing the user to focus on business logic and semantics rather than efficiency. Also, you can create your own user-defined functions (UDFs) to extend Pig Latin to do special-purpose processing.

- **Spark**: Apache Spark is a fast, in memory, parallel-processing framework that boosts the performance of big-data analytics applications. It is getting a lot of attention from the big data community because it can provide huge performance gains over MapReduce jobs. Also, it is a big data technology through which you can do streaming analytics. It works with SQL and machine learning as well.

- **Storm**: Apache Storm allows you to process large quantities of real-time data that is coming in at a high velocity. It can process up to a million records per second. It is also available as a managed service.

- **Sqoop**: Apache Sqoop is a tool to transfer bulk data to and from Hadoop and relational databases as efficiently as possible. It is used to import data from relational database management systems (RDBMS)— such as Oracle, MySQL, SQL Server, or any other structured relational database—and into the HDFS. It then does processing and/or transformation on the data using Hive or MapReduce, and then exports the data back to the RDBMS.

- **Tez**: Apache Tez is an application framework built on top of YARN to allow high-performance batch and interactive data processing. Tez is a Hindi word that means "fast," and as the name implies, it makes Hadoop jobs dramatically faster while maintaining MapReduce's scalability for petabytes of data. It is a successor of the MapReduce framework. It is more flexible and provides better performance for data-intensive processes, such as Hive and Pig.

- **ZooKeeper**: Apache ZooKeeper coordinates distributed processes on servers through shared a hierarchical name space of data registers (znode), similar to a file system. ZooKeeper is replicated over a number of hosts (called an *ensemble*) and the servers are aware of each other. Distributed applications use ZooKeeper to store and mediate updates to important configuration information.

The Advantages of HDInsight

Hadoop in HDInsight offers a number of benefits. A few of them are listed here.

- Hassle-free provisioning. Quickly builds your cluster. Takes data from Azure Blob storage and tears down the cluster when it is not needed. Use the right cluster size and hardware capacity to reduce the time for analytics and cost—as per your needs.

- Choice of using a Windows or a Linux cluster, a unique flexibility that only HDInsight provides. It runs existing Hadoop workloads without modifying single line of code.

- Another pain area in building cluster is integrating different components, such as Hive, Pig, Spark, HBase, and so forth. HDInsight provides seamless integration without your worrying about which version works with a particular Hadoop version.

- A persistent data storage option that is reliable and economical. With traditional Hadoop, the data stored in HDFS is destroyed when you tear down your cluster; but with HDInsight and Azure Blob storage, since your data is not bound to HDInsight, the same data can be fed into multiple Hadoop clusters or different applications.

- Automate cluster tasks with easy and flexible PowerShell scripts or from an Azure command-line tool.

- Cluster scaling enables you to dynamically add and remove nodes without re-creating your cluster. You can scale cluster using Azure web portal or using PowerShell/Azure command-line script.

- It can be used with the Azure virtual network to support isolation of cloud resources or hybrid scenarios where you link cloud resources with your local data center.

Summary

We live in the era of data, and it is growing at an exponential rate. Hadoop is a technology that helps you extract information from large amounts of data in a cost-effective way. HDInsight, on the other hand, is a Hadoop distribution developed by Microsoft in partnership with Hortonworks. It is easy to provision, scale, and load data in a cluster. It integrates with other Hadoop ecosystem projects seamlessly.

Next, let's dive into code and start building an HDInsight cluster.

CHAPTER 2

■ ■ ■

Provisioning an HDInsight Cluster

This chapter dives into Azure HDInsight to create an HDInsight cluster. It also goes through the different ways to provision, run, and decommission a cluster. (To do so, you need an Azure subscription. You can opt for a trial subscription for learning and testing purposes.) And finally, the chapter covers HDInsight Sandbox for local development and testing.

Microsoft Azure is a set of cloud services. One of these services is HDInsight, which is Apache Hadoop running in the cloud. HDInsight abstracts away the implementation details of installation and configuration of individual nodes. Azure Blob storage is another service offered by Azure. A blob can contain any file format; in fact, it doesn't need to know about file format at all. So you can safely dump anything from structured data to unstructured or semistructured data. HDInsight uses Blob storage as the default data store. If you store your data in Blob storage, and decommission an HDInsight cluster, data in Blob storage remains intact.

Creating an HDInsight cluster is quite easy. Open Azure portal, locate HDInsight, configure the nodes, and set permissions. You can even automate this process through PowerShell, Azure CLI, or .NET SDK if you have to do this repeatedly. The typical scenario with HDInsight and Blob storage is that you provision a cluster and run your jobs. Once the jobs are completed, you delete the cluster. With the use of Blob storage, your data remains intact in Azure for future use.

An Azure Subscription

To use any services from Azure, you need to have an Azure subscription. If you already have one, then you can skip this section; otherwise, for learning and testing purposes, you can get a free trial subscription. The trial lasts for 30 days. Note that you need to provide a valid credit card for verification, but you won't be charged until or beyond the trial period unless you switch to a Pay-As-You-Go Azure subscription. See the Frequently Asked Questions on the Azure sign-up page for more details about a trial account.

© Vinit Yadav 2017

V. Yadav, *Processing Big Data with Azure HDInsight*, DOI 10.1007/978-1-4842-2869-2_2

Creating a free trial Azure subscription:

1. To activate your trail subscription, you need to have a Microsoft account that has not already been used to sign up for a free Azure trial subscription. If you already have one, then continue to the next step; otherwise, you can get new Microsoft account by visiting `https://signup.live.com`.

2. Once you have a Microsoft account, browse to `https://azure.microsoft.com` and follow the instructions to sign up for a free trial subscription.

    a. First, you are asked to sign in with your Microsoft account, if you are not already signed in.

    b. After sign in, you are asked basic information, such as your name, email, phone number, and so forth.

    c. Then, you need to verify your identity, by phone and by credit card. Note that this information is collected only to verify your identity. You will not be charged unless you explicitly upgrade to a Pay-As-You-Go plan. After the trial period, your account is automatically deactivated if you do not upgrade to a paid account.

    d. Finally, agree to the subscription. You are now entitled to Azure's free trial subscription benefits.

You are not bound to only the trial period; if you want, you can continue to use the same Microsoft account and Azure service. You will be charged based on your usage and the type of subscription. To learn more about different services and their pricing, go to `https://azure.microsoft.com/pricing`.

Creating the First Cluster

To create a cluster, you can choose either a Windows-based or a Linux-based cluster. Both give different options for the type of cluster that you can create. Table 2-1 shows the different components that are available with different operating systems.

Table 2-1. *Cluster Types in HDInsight*

Cluster Type	Windows OS	Linux OS
Hadoop	Hadoop 2.7.0	Hadoop 2.7.3
HBase	HBase 1.1.2	HBase 1.1.2
Storm	Storm 0.10.0	Storm 1.0.1
Spark	-	Spark 2.0.0
R Server	-	R Server 9.0
Interactive Hive (Preview)	-	Interactive Hive 2.1.0
Kafka (Preview)	-	Kafka 0.10.0

Please note that Interactive Hive and Kafka are in preview at the time of writing; also, they are only available on a Linux cluster. Hortonworks is the first to provide Spark 2.0, and hence, it is available on HDInsight as well; for now, only on Linux-based clusters.

Apart from cluster types and the OS, there is one more configuration property: cluster tier. There are currently two tiers: Standard and Premium. Table 2-1 is based on what is available on a Standard tier, except R Server on Spark, which is only available on the Premium cluster tier. The premium cluster tier is still in preview and only available on a Linux cluster as of this writing.

There are multiple ways to create clusters. You can use Azure management web portal, PowerShell, Azure command-line interface (Azure CLI), or .NET SDK. The easiest is the Azure portal method, where with just a few clicks, you can get up and running, scale a cluster as needed, and customize and monitor it. In fact, if you want to create any Azure service, then Azure portal provides an easy and quick way to find those services, and then create, configure and monitor them. Table 2-2 presents all the available cluster creation methods. Choose the one that suits you best.

Table 2-2. *Cluster Creation Methods*

Cluster Creation	Browser	Command Line	REST API	SDK	Linux, Mac OS X, Unix or Windows
Azure portal	✓				✓
Azure CLI		✓			✓
Azure PowerShell		✓			✓
.NET SDK				✓	✓
cURL		✓	✓		✓
Azure Resource Manager template	✓				✓

Basic Configuration Options

No matter which method you choose to create a cluster with, you need to provide some basic configuration values. The following is a brief description of all such options.

- **Cluster name**: A unique name through which your cluster is identified. Note that the cluster name must be globally unique. At the end of the process, you are able to access the cluster by browsing to `https://{clustername}.azurehdinsight.net`.

- **Subscription name**: Choose the subscription to which you want to tie the cluster.

- **Cluster Type**: HDInsight provides six different types of cluster configurations, which are listed in Table 2-1; two are still in preview. The Hadoop-based cluster is used throughout this chapter.

    - **Operating system**: You have two options here: Windows or Linux. HDInsight is the only place where you can deploy a Hadoop cluster on Windows OS. HDInsight on Windows uses the Windows Server 2012 R2 datacenter.

    - **HDInsight version**: Identifies all the different components and the versions available on the cluster. (To learn more, go to `https://go.microsoft.com/fwLink/?LinkID=320896`)

    - **Cluster tier**: There are two tiers: Standard and Premium. The Standard tier contains all the basic yet necessary functionalities for successfully running an HDInsight cluster in the cloud. The Premium tier contains all the functionalities from the Standard tier, plus enterprise-grade functionalities, such as multiuser authentication, authorization, and auditing.

- **Credentials**: When creating an HDInsight cluster, you are asked to provide multiple user account credentials, depending on the cluster OS.

    - **HTTP/Cluster user**. This user submits jobs for admin cluster access and to access the cluster dashboard, notebook, and application HTTP/web endpoints.

    - **RDP user (Windows clusters)**. This user does the RDP connection with your cluster. When you create this user, you must set the expiry date, which cannot be longer than 90 days.

    - **SSH user (Linux clusters)**. This user does the SSH connection with your cluster. You can choose whether you want to use password-based authentication or public key–based authentication.

- **Data source**: HDInsight uses Azure Blob storage as the primary location for most data access, such as job input and logs. You can use an existing storage account or create a new one. You can use multiple storage containers with the same HDInsight cluster. Not only can you provide your own storage container, but also containers that are configured for public access.

■ **Caution** It is possible to use the same container as the primary storage for multiple HDInsight clusters—no one stops you from doing so. But this is not advisable because it may cause random problems. (More information is at `http://bit.ly/2dU4tEE`.)

- **Pricing**: On pricing the blade, you can configure the number of nodes that you want in the cluster and the size of those nodes. If you are just trying out HDInsight, then I suggest going with just a single worker node initially to keep the cost to a minimum. As you get comfortable and want to explore more scenarios, you can scale out and add more nodes. By default, there are a minimum of two head nodes.

- **Resource group**: A collection of resources that share the same life cycle, permissions, and policies. A resource group allows you to group related services into a logical unit. You can track total spending, and lock it so that no one can delete or modify it accidently. (More information is at `http://bit.ly/2dU549v`).

Creating a Cluster Using the Azure Portal

Microsoft Azure portal is the central place to provision and manage Azure resources. There are two different portals: an older portal available at `https://manage.windowsazure.com` and a newer one at `https://portal.azure.com`. This book only uses the new portal. A brief overview of the Azure portal: it is designed to provide a consistent experience no matter which service you are accessing. Once you learn to navigate and use one service, you learn to manage every other resource that Azure provides. To maintain this consistency, the Azure portal uses a blade-based UI. These blades provide a way to expose settings, actions, billing information, health condition, usage data, and much more, in a standardized way.

When you navigate to the new portal (`https://portal.azure.com`), and log in with your Microsoft account (if you are not already logged in), you are presented with a dashboard. This is not a generic dashboard; it is totally customizable. You can add new tiles or remove unnecessary tiles. You can resize them, making important ones bigger and less used ones smaller. These tiles are updated in real time to reflect up-to-date information.

The following are the steps for creating an HDInsight cluster through the Azure portal.

1. Sign in to the Azure portal (`https://portal.azure.com`).

2. Click the New button. Next, click Data + analytics, and then choose HDInsight, as shown in Figure 2-1.

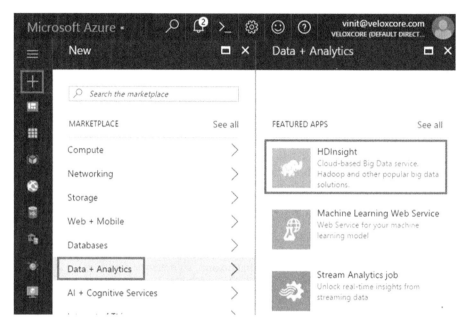

Figure 2-1. *Create new HDInsight cluster on the Azure portal*

3. Configure different cluster settings.

   a. **Cluster name**: Provide a unique name. If all rules are valid, then a green tick mark will appear at the end of it.

   b. **Cluster type**: Select Hadoop for now.

   c. **Cluster operating system**: Go with the Windows OS-based cluster.

   d. **Version**: Hadoop 2.7.3 (HDI 3.5)

   e. **Subscription**: Choose the Azure subscription that you want this cluster to be tied with.

   f. **Resource group**: Select an existing one or create a new one.

g. **Credentials**: As this is a Windows-based cluster, it can have cluster credentials. If you choose to enable the RDP connection, then it can have credentials for RDP as well, as shown in Figure 2-2. You should enable the RDP connection if you wish to get onto a head node (Windows machine).

Figure 2-2. *Windows cluster credentials*

h. **Data Source**: Create a new or select an existing storage account, and specify a primary data container as well. Figure 2-3 shows the Data Source configuration options. Also, select the Azure location where you want to create your cluster. Choose a location close to you for better performance.

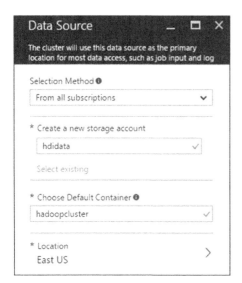

Figure 2-3. *Cluster data source*

i. **Node Pricing Tier**: Configure the number of worker nodes that the cluster will have, and both the head node size and the worker node size. Make sure that you don't create an oversized cluster unless absolutely required, because if you keep the cluster up without running any jobs, it will incur charges. These charges are based on the number of nodes and the node size that you select. (More information about node pricing is at `http://bit.ly/2dN5olv`).

j. **Optional Configuration**: You can also configure a virtual network, allowing you to create your own network in the cloud, and providing isolation and enhanced security. You can place your cluster in any of the existing virtual networks. External metastores allow you to specify an Azure SQL Database, which has Hive or Oozie metadata for your cluster. This is useful when you have to re-create a cluster every now and then. Script actions allow you to execute external scripts to customize a cluster as it is being created, or when you add a new node to the cluster. The last option is additional storage accounts. If you have data spread across multiple storage accounts, then this is the place where you can configure all such storage accounts.

You can optionally select to pin a cluster to the dashboard for quick access. Provisioning takes up to 20 minutes, depending on the options you have configured. Once the provisioning process completes, you see an icon on dashboard with the name of your cluster. Clicking it opens the cluster overview blade, which includes the URL of the cluster, the current status, and the location information.

Figure 2-4 shows the cluster provisioned just now. There is a range of settings, configurations, getting started guides, properties, and so forth, in the left sidebar. At the top, there are a few important links, discussed next.

- **Dashboard**: The central place to get a holistic view of your cluster. To get into it, you have to provide cluster credentials. Dashboard provides a browser-based Hive editor, job history, a file browser, Yarn UI, and Hadoop UI. Each provides a different functionality and easy access to all resources.

- **Remote Desktop**: Provides the RDP file, allowing you to get on to a Windows machine and the head node of your cluster. (Only available in a Windows cluster.)

- **Scale Cluster**: One of the benefits of having Hadoop in the cloud is dynamic scaling. HDInsight also allows you to change the number of worker nodes without taking the cluster down.

- **Delete**: Permanently decommissions the cluster. Note that the data stored in Blob storage isn't affected by decommissioning the cluster.

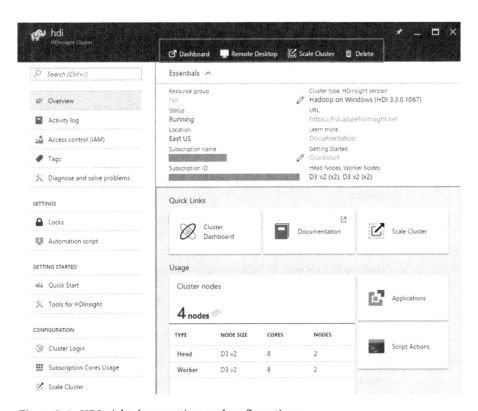

***Figure 2-4.** HDInsight cluster settings and configurations*

Connecting to a Cluster Using RDP

In the last section, you created a cluster and looked at a basic web-based management dashboard. Remote desktop (RDP) is another way to manage your Windows cluster.

To get to your Windows cluster, you must enable the RDP connection while creating the cluster or afterward. You can get the RDP file from the Azure portal by navigating to that cluster and clicking the Remote Desktop button in the header of the cluster blade. This RDP file contains information to connect to your HDInsight head node.

1. Click Remote Desktop to get the connection file for your cluster.

2. Open this file from your Windows machine, and when prompted, enter the remote desktop password.

3. Once you get to the head node, you see a few shortcuts on the desktop, as follows.

    • **Hadoop Command Line**: The Hadoop command-line shortcut provides direct access to HDFS and MapReduce, which allows you to manage the cluster and run MapReduce jobs. For example, you can run existing samples provided with your cluster by executing the following command to submit a MapReduce job.

    ```
    >hadoop jar C:\apps\dist\hadoop-2.7.1.2.3.3.1-25\
    hadoop-mapreduce-examples.jar pi 16 1000
    ```

    • **Hadoop Name Node Status**: This shortcut opens a browser and loads the Hadoop UI with a cluster summary. The same web page can be browsed using the cluster URL (https:// {clustername}.azurehdinsight.net) and navigating to the Hadoop UI menu item. From here, the user can view overall cluster status, startup progress, and logs, and browse the file system.

    • **Hadoop Service Availability**: Opens a web page that lists services, their status, and where they are running. Services included Resource Manager, Oozie, Hive, and so forth.

    • **Hadoop Yarn Status**: Provides details of jobs submitted, scheduled, running, and finished. There are many different links on it to view status of jobs, applications, nodes, and so forth.

Connecting to a Cluster Using SSH

Creating a Linux cluster is similar to Windows, except you have to provide SSH authentication instead of remote desktop credentials. SSH is a utility for logging in to Linux machines and for remotely executing commands on a remote machine. If you are on Linux, Mac OS X, or Unix, then you already have the SSH tool on your machine; however, if you are using a Windows client, then you need to use PuTTY.

When creating a Linux cluster, you can choose between password-based authentication and public-private key–based authentication. A password is just a string, whereas a public key is a cryptographic key pair to uniquely identify you. While password-based authentication seems simple to use, key-based is more secure. To generate a public-private key pair, you need to use the PuTTYGen program (download it from http://bit.ly/1jsQjnt). You have to provide your public key while creating a Linux-based cluster. And when connecting to it by SSH, you have to provide your private key. If you lose your private key, then you won't be able to connect to your name node.

The following are the steps to connect to your Linux cluster from a Windows machine.

1. Open PuTTY and enter the Host Name as **{clustername}-ssh. azurehdinsight.net** (for a Windows client). Keep the rest of the settings as they are.

2. Configure PuTTY based on the authentication type that you select. For key-based authentication type, navigate to Connection, open SSH, and select Auth.

3. Under Options controlling SSH authentication, browse to the private key file (PuTTY private key file *.ppk).

4. If this is a first-time connection, then there is a security alert, which is normal. Click Yes to save the server's RSA2 key in your cache.

5. Once the command prompt opens, you need to provide your SSH username (and password if configured so). Soon the SSH connection is established with the head node server.

To monitor cluster activity, there is Ambari Views. You can find a shortcut for the same in the Linux cluster's Overview blade under the Quick Links section. Ambari shows a complete summary of the cluster, including HDFS Disk usage, memory, CPU and network usage, current cluster load, and more.

■ **Warning** HDInsight clusters billing is pro-rated per minute, whether you are using them or not. Please be sure to delete your cluster after you have finished using it.

Creating a Cluster Using PowerShell

Azure PowerShell is a module that provides cmdlets to manage Azure resources from within Windows PowerShell. There are multiple ways to install the Azure PowerShell module; the easiest are the Microsoft Web Platform Installer or the PowerShell Gallery. All available installation options are at http://bit.ly/2bF85UH. You can verify installation by opening Windows PowerShell and executing the "Get-Module -ListAvailable -Name Azure" command. As shown in Figure 2-5, the command returns the currently installed version of the Azure PowerShell module.

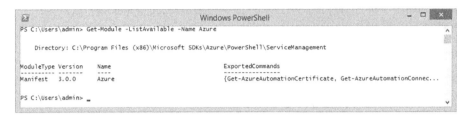

Figure 2-5. *Azure PowerShell module version*

Now that you have PowerShell set up correctly, let's create an Azure HDInsight cluster. First, log in to your Azure subscription. Execute the "Login-AzureRmAccount" command on the PowerShell console. This opens a web browser. Once you authenticate with a valid Azure subscription, PowerShell shows your subscription and a successful login, as shown in Figure 2-6.

Figure 2-6. *Log in to Azure from Windows PowerShell*

If you happen to have more than one Azure subscription and you want to change from the selected default, then use the "Add-AzureRmAccount" command to add another account. The complete Azure cmdlet reference can be found at http://bit.ly/2dMxlMo.

With an Azure resource group, once you have PowerShell configured and you have logged in to your account, you can use it to provision, modify, or delete any service that Azure offers. To create an HDInsight cluster, you need to have a resource group storage account. To create a resource group, use the following command.

```
New-AzureRmResourceGroup -Name hdi -Location eastus
```

To find all available locations, use the "Get-AzureRmLocation" command. To view all the available resource group names, use the "Get-AzureRmResourceGroup" command.

On a default storage account, HDInsight uses a Blob storage account to store data. The following command creates a new storage account.

```
New-AzureRmStorageAccount -ResourceGroupName hdi -Name hdidata -SkuName
Standard_LRS -Location eastus -Kind storage
```

Everything in the preceding command is self-explanatory, except LRS. LRS is *locally redundant storage*. There are five types of storage replication strategies in Azure:

- Locally redundant storage (Standard_LRS)

- Zone-redundant storage (Standard_ZRS)

- Geo-redundant storage (Standard_GRS)

- Read-access geo-redundant storage (Standard_RAGRS)

- Premium locally redundant storage (Premium_LRS)

More information about storage accounts is in Chapter 3.

■ **Note** The storage account must be collocated with the HDInsight cluster in the data center.

After creating a storage account, you need to get an account key. The following command gets a key for a newly created or an existing storage account.

```
-- Lists Storage accounts
Get-AzureRmStorageAccount
-- Shows a Storage account
Get-AzureRmStorageAccount -AccountName "<Storage Account Name>"
-ResourceGroupName "<Resource Group Name>"
-- Lists the keys for a Storage account
Get-AzureRmStorageAccountKey -ResourceGroupName "<Resource Group Name>"
-Name "<Storage Account Name>" | Format-List KeyName,Value
```

After you are done with the resource group and the storage account, use the following command to create an HDInsight cluster.

```
New-AzureRmHDInsightCluster [-Location] <String> [-ResourceGroupName]
<String>
        [-ClusterName] <String>  [-ClusterSizeInNodes] <Int32>
        [-HttpCredential] <PSCredential>
        [-DefaultStorageAccountName] <String>]
        [-DefaultStorageAccountKey] <String>
        [-DefaultStorageContainer <String>]
        [-HeadNodeSize <String>] [-WorkerNodeSize <String>]
        [-ClusterType <String>]
        [-OSType <OSType>] [-ClusterTier <Tier>]
        [-SshCredential <PSCredential>] [-SshPublicKey <String>]
        [-RdpCredential <PSCredential>] [-RdpAccessExpiry <DateTime>]
```

The preceding code block is a stripped-down version of all available options when creating a cluster. The following is a sample PowerShell script.

```
$clusterHttpCreds = Get-Credential
$clusterSSHCreds = Get-Credential
New-AzureRmHDInsightCluster -Location eastus -ResourceGroupName hdidata
-ClusterName hdi
        -ClusterSizeInNodes 2 -HttpCredential $clusterHttpCreds
        -DefaultStorageAccountName "hdistorage.blob.core.windows.net"
        -DefaultStorageAccountKey "mXzSxbPsE2...oS9TSUw=="
        -DefaultStorageContainer hdicontainer -HeadNodeSize Standard_D3_V2
        -WorkerNodeSize Standard_D3_V2 -ClusterType Hadoop -OSType Linux
        -ClusterTier Standard -SshCredential $clusterSSHCreds
```

Note that this script assumes that you have already created the "hdidata" resource group and the "hdistorage" storage account as well. Also, note that defaultStorageAccountKey is not a complete key; you should replace it with yours before executing. To learn all the available node sizes, execute the "Get-AzureRmVMSize -Location eastus" command.

After the data processing task is complete and the cluster is no longer required, you can delete it using following command.

```
Remove-AzureRmHDInsightCluster -ClusterName "hdi"
```

Creating a Cluster Using an Azure Command-Line Interface

The Azure command-line interface has open source shell-based commands for managing resources in Microsoft Azure. Azure CLI can be installed using the node package manager (npm), installer file, or the Docker container. The simplest way is to install it using an installer: Windows installer (http://aka.ms/webpi-azure-cli) and Mac OS X installer (http://aka.ms/mac-azure-cli).

To install using npm, execute the "npm install -g azure-cli" command for a Windows machine, and on Linux, use "sudo npm install -g azure-cli". Please note that for npm to work, you need to install Node.js first. The third and last option is to install Azure CLI is Docker. You need to set up your machine as a Docker host. Run the "docker run -it microsoft/azure-cli" command (on Linux you need to use sudo). You can verify installation by running the "Azure" command on Azure CLI.

After configuration, Azure CLI is almost similar to PowerShell. Try to log in using the "Azure login" command. Just like PowerShell, this command provides you a token and a URL (https://aka.ms/devicelogin). You have to open a browser, navigate to the provided URL, and enter the given token. This logs you in to Azure CLI automatically. Also, if you have only one subscription, it selects it for all operations. You can change it by using the "`azure account set <subscriptionNameOrId>`" command where you need to provide a subscription name or id. Go to `http://bit.ly/2dE5OYu` for complete command references.

With an Azure resource group, once you have CLI configured, and you have logged into your account, you can use it to provision, modify, or delete any service that Azure offers. To create an HDInsight cluster, you need to have a resource group storage account. To create a resource group, use the following command.

```
azure group create -n "<Resource Group Name>" -l "<Azure Location>"
```

To find all available locations use the "`azure location list`" command. To view all available resource group names, use the "`azure group list`" command.

On a default storage account, HDInsight uses a Blob storage account to store data. The following command creates a new storage account.

```
azure storage account create "<NAME>" -g "<Resource Group Name>" -l "<Azure
Location>" --sku-name LRS --kind Storage
```

Everything in this command is self-explanatory, except LRS. LRS is locally redundant storage. The following are four types of storage replication strategy in Azure:

- Locally redundant storage (LRS)

- Zone-redundant storage (ZRS)

- Geo-redundant storage (GRS)

- Read-access geo-redundant storage (RA-GRS)

More information about storage accounts is in Chapter 3.

After creating a storage account, you need to get an account key. The following command helps you get a key from a newly created or an existing storage account.

```
-- Lists Storage accounts
azure storage account list
-- Shows a Storage account
azure storage account show "<Storage Account Name>" -g "<Resource Group
Name>"
-- Lists the keys for a Storage account
azure storage account keys list "<Storage Account Name>" -g "<Resource Group
Name>"
```

After you are done with the resource group and the storage account, use the following command to create an HDInsight cluster.

```
azure hdinsight cluster create
-g <Resource Group Name> -c <HDInsight Cluster Name> -l <Location>
--osType <Windows | Linux> --version <Cluster Version>
--clusterType <Hadoop | HBase | Spark | Storm>
--workerNodeCount 2 --headNodeSize Large --workerNodeSize Large
--defaultStorageAccountName <Azure Storage Account Name>.blob.core.windows.net
--defaultStorageAccountKey "<Default Storage Account Key>"
--defaultStorageContainer <Default Blob Storage Container>
--userName admin --password "<HTTP User Password>"
--rdpUserName <RDP Username> --rdpPassword "<RDP User Password>"
--rdpAccessExpiry "<ExpiryDate>"
```

The preceding code block is a stripped-down version of all the available options when creating a cluster. The following is a sample command.

```
azure hdinsight cluster create -g hdidata -c hdiClusterName -l eastus
--osType Windows --version 3.3 --clusterType Hadoop --workerNodeCount 2
--headNodeSize Large --workerNodeSize Large --defaultStorageAccountName
hdistorage.blob.core.windows.net --defaultStorageAccountKey lghpx6p94...
WXrvCsIb1atLIyhg== --defaultStorageContainer hdi1 --userName admin
--password AbCdE@12345 --rdpUserName rdpuser --rdpPassword AbCdE@98765
--rdpAccessExpiry "01/01/2017"
```

Note that the preceding command assumes that you have the "hdidata" resource group already created, as well as an "hdistorage" storage account. Also note that defaultStorageAccountKey is not a complete key; you should replace it with yours before executing.

After the data processing task is complete and the cluster is no longer required, you can delete it by using the following command.

```
azure hdinsight cluster delete <CLUSTERNAME>
```

Creating a Cluster Using .NET SDK

Microsoft provides .NET SDK to work with HDInsight clusters so that you can integrate it in .NET applications. Just like Azure CLI and PowerShell, if you learn to work with any one type of resource, you can literally work with almost any service Azure offers through .NET SDK. Next, let's set up authentication and then create an HDInsight cluster.

To start, open Visual Studio (2012 or higher) and follow these instructions:

1. Create a new C# console application. (VB.NET also works, if you prefer it.)

2. To get SDK components, run the following Nuget commands in the Nuget Package Management console. Please note that the resources Nuget is still in preview at the time of writing.

```
Install-Package Microsoft.Rest.ClientRuntime.Azure.Authentication
Install-Package Microsoft.Azure.Management.Resources -Pre
Install-Package Microsoft.Azure.Management.Storage
Install-Package Microsoft.Azure.Management.HDInsight
```

3. Once all packages are installed, you can create an HDInsight cluster. But as in the previous section, you will first create a resource group and a storage account. Before that, let's have a look at authentication. The following code snipet shows how to log in to your Azure subscription.

```
static TokenCloudCredentials Authenticate(string TenantId, string
ClientId, string SubscriptionId)
{
    var authContext = new AuthenticationContext
    ("https://login.microsoftonline.com/"
        + TenantId);

    var tokenAuthResult = authContext.AcquireToken
    ("https://management.core.windows.net/",
        ClientId,
        new Uri("urn:ietf:wg:oauth:2.0:oob"),
        PromptBehavior.Auto,
        UserIdentifier.AnyUser);

    return new TokenCloudCredentials(SubscriptionId,
    tokenAuthResult.AccessToken);
}
```

4. The preceding method requires few ids. TenantId and SubscriptionId can be found in portal.azure.com. After login, go to Help and select Show Diagnostics. This opens a new browser window with a lot of information. You can find both TenantId and SubscriptionId in it. About ClientId: I use the PowerShell client's GUID, which is used for interactive logins. If you have your own application attached with Azure Active Directory, then you can use the client id of that application as well.

5. Once you execute the preceding method, and if you are already logged in to the Azure portal from Internet Explorer, then you automatically get an authentication token; otherwise, you have to enter your credentials in the browser window, opened by the code, to get an authentication token. The preceding method waits for execution on the second line until it gets an authentication token.

6. After the code receives the token, you can work with different services. The following are a few methods to work with resource groups. Note that all of these methods require credentials passed as a parameter, which you received from the authentication method.

```
static async Task CreateResourceGroup(string resourceGroupName,
TokenCloudCredentials credentials)
{
    var resourceClient =
        new Microsoft.Azure.Management.Resources.ResourceManagement
        Client(credentials);

    Microsoft.Azure.Management.Resources.Models.
    ResourceGroupCreateOrUpdateResult result =
        await resourceClient.ResourceGroups.CreateOrUpdateAsync(
            resourceGroupName,
            new Microsoft.Azure.Management.Resources.Models.
            ResourceGroup(location: "eastus"));

    Console.WriteLine($"Resource group creation result: {result.
    StatusCode.ToString()}");
}

static async Task ListResourceGroups(TokenCloudCredentials
credentials)
{
    var resourceClient =
        new Microsoft.Azure.Management.Resources.ResourceManagement
        Client(credentials);

    var resources =
        await resourceClient.ResourceGroups.ListAsync(
            new Microsoft.Azure.Management.Resources.Models.
            ResourceGroupListParameters());

    foreach (Microsoft.Azure.Management.Resources.Models.
    ResourceGroupExtended group
        in resources.ResourceGroups)
```

```
    {
        Console.WriteLine($"Name: {group.Name}, Location:
    {group.Location}");
    }
}

static async Task DeleteResourceGroup(string resourceGroupName,
TokenCloudCredentials credentials)
{
    var resourceClient =
        new Microsoft.Azure.Management.Resources.ResourceManagement
        Client(credentials);

    AzureOperationResponse result =
        await resourceClient.ResourceGroups.DeleteAsync
        (resourceGroupName);

    Console.WriteLine($"Resource group deletion result: {result.
    StatusCode.ToString()}");
}
```

7. To work with a storage account, the following are the methods
   code to create, list, or delete. Note that all of these methods
   require credentials, as received from the authentication
   method.

```
static async Task CreateStorageAccount(string resourceGroupName,
string accountName,
    TokenCloudCredentials credentials)
{
    var storageClient =
        new Microsoft.Azure.Management.Storage.StorageManagement
        Client(credentials);

    var result = await storageClient.StorageAccounts.CreateAsync(
        resourceGroupName: resourceGroupName,
        accountName: accountName,
        parameters: new Microsoft.Azure.Management.Storage.
        Models.StorageAccountCreateParameters
                (accountType: Microsoft.Azure.Management.
                Storage.Models.AccountType.StandardLRS,
                location: "eastus"));

    Console.WriteLine($"Storage Account creation result: {result.
    StatusCode.ToString()}");
}
```

```csharp
static async Task ListStorageAccounts(TokenCloudCredentials
credentials)
{
    var storageClient =
        new Microsoft.Azure.Management.Storage.StorageManagement
        Client(credentials);

    var resources = await storageClient.StorageAccounts.ListAsync();

    foreach (Microsoft.Azure.Management.Storage.Models.
    StorageAccount storageAccount
        in resources.StorageAccounts)
    {
        Console.WriteLine($"Name: {storageAccount.Name},
        Location: {storageAccount.Location}");
    }
}

static async Task DeleteStorageAccount(string resourceGroupName,
string accountName,
    TokenCloudCredentials credentials)
{
    var storageClient =
        new Microsoft.Azure.Management.Storage.StorageManagement
        Client(credentials);

    AzureOperationResponse result =
        await storageClient.StorageAccounts.
        DeleteAsync(resourceGroupName, accountName);

    Console.WriteLine($"Storage Account deletion result: {result.
    StatusCode.ToString()}");
}

static async Task<Microsoft.Azure.Management.Storage.Models.
StorageAccountKeys>
    GetStorageAccountKey(string resourceGroupName, string
    accountName, TokenCloudCredentials credentials)
{
    var storageClient =
        new Microsoft.Azure.Management.Storage.StorageManagementC
        lient(credentials);

    var result = await storageClient.StorageAccounts.ListKeysAsync
    (resourceGroupName, accountName);

    return result.StorageAccountKeys;
}
```

```
static async Task<string> GetBlobLocation(string
resourceGroupName, string accountName,
    TokenCloudCredentials credentials)
{
    var storageClient =
        new Microsoft.Azure.Management.Storage.StorageManagement
        Client(credentials);

    var result = await storageClient.StorageAccounts.
    GetPropertiesAsync(
        resourceGroupName, accountName);

    return result.StorageAccount.PrimaryEndpoints.Blob.Host;
}
```

8. Finally, after you have a resource group and a storage account, you can create an HDInsight cluster. The following main method uses previous code. Please note that this code doesn't gracefully handle failures (such as a down Internet connection). It is used only for demonstration purposes.

```
static void Main(string[] args)
{
    string TenantId = "<Tenant Id>";
    string SubscriptionId = "<Azure Subscription Id>";

    // This is the GUID for the PowerShell client. Used for
    interactive logins in this example.
    string ClientId = "1950a258-227b-4e31-a9cf-717495945fc2";

    string clusterName = "HDICluster";
    string resourceGroupName = "<Resource Group Name>";
    string storageAccountName = "<Storage Account Name>";

    // Get authentication token
    var credentials = Authenticate(TenantId, ClientId,
    SubscriptionId);

    // Working with Resource group
    CreateResourceGroup(resourceGroupName, credentials).Wait();

    // Working with Storage Account
    CreateStorageAccount(resourceGroupName, storageAccountName,
    credentials).Wait();
    Microsoft.Azure.Management.Storage.Models.StorageAccountKeys
    storageAccountKeys =
        GetStorageAccountKey(resourceGroupName,
        storageAccountName, credentials).Result;
```

```csharp
    string blobContainer = GetBlobLocation(resourceGroupName,
    storageAccountName,
        credentials).Result;

    System.Console.WriteLine("Creating a cluster. The process may
    take 10 to 20 minutes...");

    // Get an HDInsight management client
    HDInsightManagementClient hdiManagementClient = new HDInsight
    ManagementClient(credentials);

    // Set parameters for the new cluster
    var parameters = new ClusterCreateParameters
    {
        Location = "EAST US",
        OSType = OSType.Windows,
        Version = "3.3",
        ClusterType = "Hadoop",
        ClusterSizeInNodes = 2,
        HeadNodeSize = "Large",
        WorkerNodeSize = "Large",
        DefaultStorageAccountName = blobContainer,
        DefaultStorageAccountKey = storageAccountKeys.Key1,
        DefaultStorageContainer = "hdicontainer",
        UserName = "admin",
        Password = "AbCdE@12345",
        RdpUsername = "rdpuser",
        RdpPassword = "AbCdE@12345",
        RdpAccessExpiry = DateTime.Now.AddMonths(2)
    };

    // Create the cluster
    hdiManagementClient.Clusters.Create(resourceGroupName,
    clusterName, parameters);

    System.Console.WriteLine("The cluster has been created. Press
    any key to delete cluster...");
    System.Console.ReadLine();

    // Delete the cluster (if required, otherwise remove the
    below code)
    hdiManagementClient.Clusters.Delete(resourceGroupName,
    clusterName);
}
```

To run the whole app, just create a C# console app and paste all the code in a Program.cs file, change all the missing values (TenantId, SubscriberId, etc.), and press F5. You will have created a cluster.

The Resource Manager Template

The Azure Resource Manager template makes it even easier to create an HDInsight cluster. The template is just a JSON file with all the configuration parameters and dependent components, such as a storage account or a SQL database (for Apache Sqoop), specified in it. This single file can contain any number of services that you want to deploy, working in single coordinated operation, making it a powerful way to provision services in Azure. Also, since it is just a JSON file, you don't have to know PowerShell, Azure CLI, or .NET code to create it. The template defines the resources that need to be created with parameters to input values for different environments.

If you don't want to create it manually, then there is a way around it. Log in into `https://portal.azure.com` and configure the HDInsight cluster as described in the "Creating Cluster Using the Azure Portal" section. Beside the Create button, there is an "Automation options" link. Clicking it opens a new blade with an HDInsight template and parameter file. PowerShell, Azure CLI, .NET, and Ruby also execute this template.

HDInsight in a Sandbox Environment

To do development, it is not feasible to keep an HDInsight cluster running on Azure. It would be neither cost effective nor easy to do development on a remote cluster. As developers, we always prefer our own local environment, where we can quickly test the app without any dependency. To help with such scenarios, Hortonworks provides the Sandbox environment. You can have either a Linux-based cluster or a Windows-based cluster. A Linux-based sandbox is available with the latest HDP 2.5 (Hortonworks Data Platform), whereas the Windows-based sandbox is currently available for HDP 2.3.4. The next section demonstrates how to set up both of these environments.

Hadoop on a Virtual Machine

With Hortonworks Sandbox, you can install and run Hadoop locally to learn about Hadoop, HDFS, and job submissions, as well as its ecosystem components. Make sure that you have a machine with a 64-bit multicode CPU that supports virtualization, has at least 8GB of RAM (the more, the better), a browser (Chrome 25+, IE9+, or Safari 6+), and the latest version of VirtualBox (4.2 or higher from `www.virtualbox.org`) installed. the following steps are for Windows, but are the same for Mac OS or Linux.

1. Hortonworks Sandbox is available for VirtualBox, VMware, and as a Docker image. You can choose any of these options and get the same result. Let's continue with the VirtualBox option. The Hortonworks Sandbox is available as a VirtualBox appliance. To download, go to `http://hortonworks.com/downloads/` and download it for VirtualBox, as shown in Figure 2-7.

Hortonworks Sandbox on a VM

HDP® 2.5 on Hortonworks Sandbox

Figure 2-7. *Hortonworks Sandbox options*

2. After downloading the appliance file (*.ova), it needs to be imported into VirtualBox. Open the Oracle VirtualBox and go to File ➤ Import Appliance. First, you need to provide the appliance file. After selecting an .ova file and clicking Next, you see all the configuration details of the new VM that is to be created (change it as you will, but for now leave it as it is). On the same Appliance settings page, click Import, which starts the import process. Soon a new VM named Hortonworks Sandbox appears in VirtualBox.

3. To run it, just select the VM node and click the green start button. Once the virtual machine has started, open the web browser and go to http://127.0.0.1:8888. this opens a page showing two options: Launch Dashboard and Quick Links. The first one opens the Ambari dashboard and the other one lists the different services and explains how to access them. The Quick Links button opens the web page (http://127.0.0.1:8888/splash2.html) shown in Figure 2-8, which explains how to access services like Ambari, Atlas, Falcon, Ranger, Zeppelin, and SSH Client.

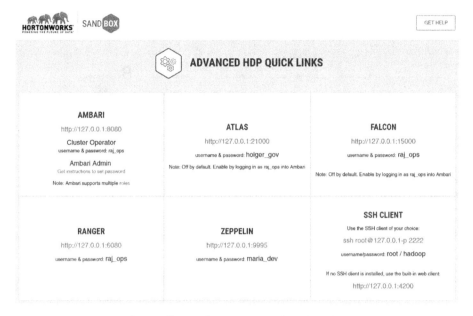

Figure 2-8. *Hortonworks Sandbox with HDP 2.5 Quick Links web page*

Apache Ambari simplifies the management and monitoring of a Hadoop cluster from within a web browser. Ambari also provides REST APIs so that the developer can integrate these capabilities in their custom apps. Figure 2-9 shows the Ambari dashboard from the Hortonworks Sandbox environment. Go ahead and play around with the Dashboard, Services, Hosts, Alerts, and Admin pages. You will be surprised how easy it is to work with Ambari.

Figure 2-9. *Hortonworks Sandbox: Ambari Web UI*

4. To connect with a cluster through SSH, you can view details on the Quick Links page. If you are on a Windows host and don't want to download the SSH client, then just navigate to http://127.0.0.1:4200 in your browser to connect using SSH. (If you are trying to connect for first time, then it asks you to change your password.) Once logged in, try running the "hive" command, which connect you to the Hive shell.

5. Once connected to the Hive shell, run the Show tables; command to view the available sample tables. HDP 2.5 shows your two tables: sample_07 and sample_08.

6. Run the select * from sample_07 limit 10; command to retrieve the first ten rows from the sample_07 table.

You can also use Visual Studio to connect to the HDInsight cluster. Open Visual Studio or from Server Explorer ➤ Azure ➤ HDInsight. Right-click the HDInsight node and select Connect to HDInsight Emulator. This opens what's shown in Figure 2-10, where you can enter information about the Sandbox environment to connect with it.

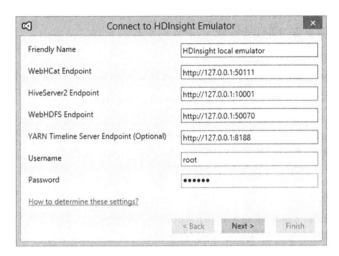

Figure 2-10. *Visual Studio HDInsight Emulator connection parameters*

After Visual Studio connects to the Sandbox environment, right-click the emulator node. It shows the number of options, such as write a Hive query, view jobs, or browse HDFS UI. Without leaving Visual Studio, you can submit a Hive query directly. (If you want, you can try running the same query from the last section in here as well.)

Hadoop on Windows

The Hortonworks Data Platform is the only Hadoop distribution available on Windows host machines. They have provided a way to create local cluster on Windows machines too. Following are the steps to configure a Hadoop cluster on a Windows machine.

■ **Note**   HDP for Windows is available as a single-node cluster and for evaluation purposes only.

Preparing the Host Machine

Before beginning cluster installation, the host machine should meet the following hardware and software requirements.

- Host OS should be one of the following:

  - Windows Server 2008 R2 (64-bit)

  - Windows Server 2012 (64-bit)

  - Windows Server 2012 R1/R2 (64-bit)

- All of the following software should be installed and configured correctly:

  - Microsoft Visual C++ 2010 Redistributable Package (64-bit) (Download from `http://bit.ly/1dzHJyx`.)

  - Java JDK 1.7.x

  - Python 2.7.x

- Disable your Windows firewall so that all the ports are available for use. (Instructions are at `http://bit.ly/2eqLuR4`.)

Installing and Configuring Java JDK

1. Download Java JDK (v1.7.x or higher) and install it in the `C:\Java` directory. After installation, verify that you have the JDK folder located inside the Java folder.

2. Add the JAVA_HOME environment variable:

    a. Open Run and type **SystemPropertiesAdvanced. exe**, which opens System Properties, and then click Environment Variables.

    b. In the Environment Variables window, click New in the System variables section and add the JAVA_HOME environment variable pointing to `c:\Java\jdk1.7.0_51`. Verify that it saved successfully by opening a command prompt and executing `Echo %JAVA_HOME%`.

    c. Change Path system variable to include Java bin folder, to do that find Path variable in list and click on Edit. Append `c:\Java\jdk1.7.0_51\bin` at the end of existing Path variable value.

Installing and configuring Python 2.7.x

1. Download Python and install it in a directory C:\Python27. (Download from www.python.org/downloads/.)

2. After download and install, add C:\Python27 to the path system variable.

3. Verify installation by opening a command line or PowerShell and executing python -V Python 2.7.x (where x is the version that you downloaded).

Download and Install HDP for Windows

1. Download the HDP for Windows installer from http://public-repo-1.hortonworks.com/HDP-Win/2.3/2.3.4.0/hdp-2.3.4.0.zip.

2. Open a command line and write the following command to launch HDP for Windows installer.

   ```
   Runas /user:administrator "cmd /C msiexe /lv c:\hdplog.txt
   /I {PATH_TO_MSI} MSIUSEREALADMINDETECTION=1"
   ```

3. This opens the HDP setup window, as shown in Figure 2-11. Make sure that it is configured correctly before starting installation. To configure, use the following steps:

   a. Enter your Hadoop user password. A stronger password is preferred.

   b. Enter the Hive and Oozie database details.

   c. Modify the default DB flavor selection from MSSQL to Derby database.

   d. Click the Additional Components tab. Uncheck Apache Ranger from the list of available additional components.

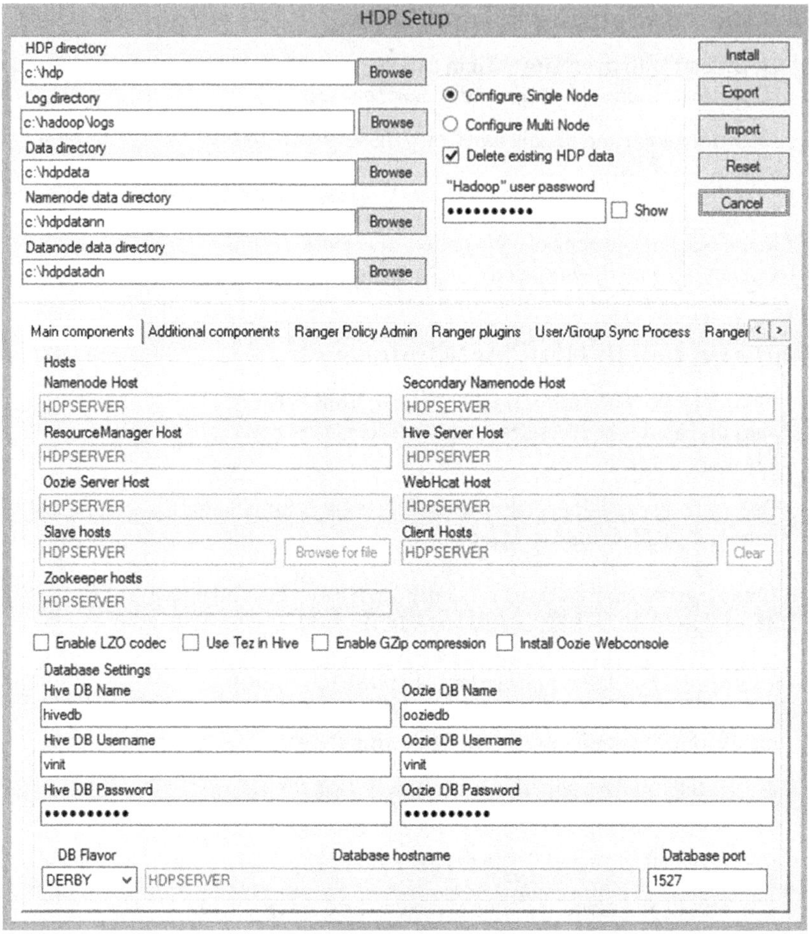

Figure 2-11. *HDP Windows setup*

4. Add a password on the Additional Components screen and hit
Install. It may take up to 20 minutes.

After installation finishes, you have a Hadoop cluster ready to explore. Similar to the
RDP machine in Azure, you also get a few shortcuts on your desktop. You can start/stop
all the services using the Hadoop command line. To start HDP services, open the Hadoop
command line and navigate to the HDP install directory (if you followed the same paths
as in Figure 2-11, then it is `c:\hdp`). Enter the `start_local_hdp_services.cmd` command.
That is all that is needed to set up a local Windows cluster.

Summary

Azure HDInsight offers many different ways to provision and maintain a Hadoop cluster. Whether you are a developer, a data scientist, or a DevOps engineer, you feel right at home with HDInsight's provisioning methods. Hortonworks Data Platform (HDP) complements what Microsoft offers, and gives you a local environment on both Linux and Windows. Configuring a local environment is easy because it's cloud counterpart and tooling support in Visual Studio makes developer's life simple by shortening different integration pain points with services like Hive and Pig. By now, you have a cluster are ready to execute jobs on it. Let's look at the different ways in which you can execute MapReduce jobs on an HDInsight cluster.

■ ■ ■

## CHAPTER 3

Working with Data in HDInsight

Azure Blob storage is the default and preferred way to store data in HDInsight. HDInsight supports the Hadoop distributed file system (HDFS) as well as Azure Blob storage for storing data. This chapter covers uploading data to Blob storage and executing MapReduce jobs on it. It starts with different command-line utilities to upload data and looks at a couple of graphical clients. You'll create your first MapReduce job and execute it using PowerShell. Also, you'll look at .NET SDK to create and execute job on HDInsight. And finally, you'll learn about Avro serialization.

Azure Blob Storage

Azure Blob storage is a general-purpose storage solution that can store structured or unstructured data and integrate seamlessly with HDInsight. It is the default storage used by HDInsight. Keeping data in Blob storage facilitates the safe deletion of a cluster without losing any data. Using the HDFS interface, Hadoop components can directly work on data stored in Blob storage. HDInsight has a notion of the default file system, which implies a default schema and default authority, also resolving relative paths to the default storage account. You have to provide default an Azure Blob storage container while creating HDInsight a cluster, which you saw in the previous chapter's discussion of the HDInsight cluster creation process.

The blob service contains three components: the storage account, the container, and Blob storage. There are two types of storage accounts: general purpose, which includes tables, queues, files, blobs, and Azure virtual machine disks, and the Blob storage account, which stores unstructured data as objects (blobs). A container is a logical grouping of blobs (a blob is a file of any type or size). To visualize a storage account, consider Figure 3-1, where there is a storage account HDI, which has two containers and each of them stores files as blobs.

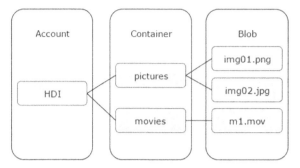

Figure 3-1. *Blob service components*

The Benefits of Blob Storage

Hadoop has a data locality principle, which states that data should reside as close to the compute node as possible to reduce the data movement and enable faster processing. With HDInsight, data is stored in Blob storage; not within the compute node. But performance is compensated by having data near to the HDInsight cluster, in the same Azure region, and a very high-speed network connecting them. The following are a few of the benefits of using Azure Blob storage for storing data vs. storing the data on individual compute nodes.

- Data stored inside HDFS can only be consumed by using HDFS APIs, while data inside Blob storage can be accessed through REST APIs too. Hence, a larger set of applications and components can use this data.

- HDFS keeps redundant files inside data nodes; whereas Blob storage can have georeplication. This also improves write operations, as HDFS has to finish all writes (if the replication factor is 3, then it writes three copies of data) before marking a write operation complete; on the other hand, Blob storage has to just write it once. Azure takes care of replication subsequently.

- Keeping data in DFS is pricier than using Blob storage for the same data because the cost of a compute cluster is higher than a Blob storage container.

- It is safe to delete a cluster without losing any data with Blob storage; whereas HDFS is destroyed along with the cluster.

- HDFS scale-out requires adding new nodes, even if you don't need more computation power, which is an overhead; whereas Azure Blob storage provides elastic scaling capabilities.

■ **Note** You can use most of the HDFS commands (i.e., ls, copyFromLocal, and mkdir) with HDInsight and Azure Blob storage. Only a few commands specific to HDFS implementation (DFS-related, i.e., fschk or dfadmin) will show different behavior in Azure Blob storage.

Blob storage is referred to as Windows Azure Storage Blob (WASB). WASB is an extension built on top of the HDFS APIs. To access data stored in a Blob storage container, the URI is as follows.

```
wasb[s]://<containername>@<accountname>.blob.core.windows.net/<path>
```

For example, to access a file named *example/sample.txt*, located at the root of a container named hdicontainer, inside the hdistorage storage account, the following URI schema can be used.

```
wasb[s]://hdicontainer@hdistorage.blob.core.windows.net/example/sample.txt
wasb[s]://example/sample.txt
/example/sample.txt
```

Here WASB uses SSL certificates for secure connection and improved security. WASB provides a layer of abstraction over the storage, which enables you to persist data even if the cluster is decommissioned. Also, it provides access to multiple applications and clusters at the same time, providing overall increased flexibility.

Along with Blob storage, you also get HDFS storage in the cluster nodes. That means WASB (virtually) and HDFS reside side by side in your node. To access the HDFS storage available inside your cluster nodes, use the following URI.

```
Hdfs://<namenodehost>:<port>/<path>
```

Figure 3-2 shows the HDInsight storage architecture, where DFS and WASB are shown side by side, and internally, WASB uses Azure Blob storage containers.

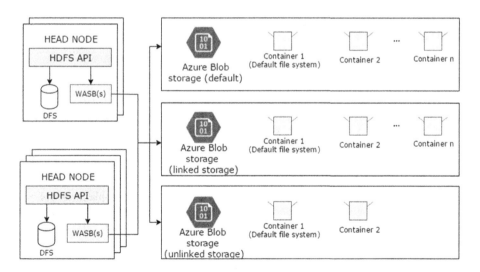

Figure 3-2. *HDInsight storage architecture*

Uploading Data

Azure Blob storage stores data as key-value pairs. Also, it is a flat structure and does not have directory hierarchy. However, you can add the / character to a key, making it look like directory. Many client tools (i.e., Azure Storage Explorer, etc.) consider it logical grouping and present it as a hierarchical structure. For example, a blob's key can be example/sample.txt. Here, no actual example directory exists, and the / character has no special meaning; still, it appears as a file path.

There are many ways to upload data to Blob storage. Table 3-1 lists many command-line utilities. Table 3-2 lists graphical clients that help you upload data.

Table 3-1. *Command-Line Utilities*

Tool	Linux OS	Mac OS	Windows
Azure Command-Line Interface	✓	✓	✓
Azure PowerShell			✓
AzCopy			✓
Hadoop command-line tool	✓	✓	✓

Table 3-2. *Graphical Clients*

Tool	Linux OS	Mac OS	Windows
Microsoft Visual Studio Tools for HDInsight			✓
Azure Storage Explorer	✓	✓	✓
Cloud Storage Studio 2			✓
CloudXplorer			✓

In addition to these tools, you can upload/download files programmatically with .NET SDK and integrate the process in custom applications. The next section covers cross-platform Azure CLI, Microsoft Azure Storage Explorer, and .NET SDK.

Using Azure Command-Line Interface

As discussed in Chapter 2, Azure CLI (command-line interface) is a cross-platform tool for managing Azure services. Installation instructions are in Chapter 2 in the "Creating a Cluster Using the Azure Command-Line Interface" section or at http://bit.ly/2jKtQXe. Follow these steps to upload data to Blob storage using Azure CLI.

1. Open Azure CLI and log in to your Azure subscription, using the azure login command. This shows a code and a URL: https://aka.ms/devicelogin.

2. Open a browser and navigate to the provided URL. Enter the code and Azure subscription credentials when prompted. Afterward, the browser shows that you are signed in to the Microsoft Azure cross-platform command-line interface. Switch to Azure CLI and you see that you are now logged in.

3. To upload or download files requires a storage account name, a primary key, and a container name. To see all available storage accounts, execute the following command.

```
azure storage account list
```

4. Take any one of the storage accounts and find the primary key for the same. To do so, execute the following account with the storage account name and its resource group name.

```
azure storage account keys list <storage_account_name>
-g <group_name>
```

■ **Note** Azure Blob storage account has two keys: primary and secondary. You can access storage with only one key. Then why are there two keys? It is for rolling key changes. Assume that you want to change one of the keys because it was leaked somehow, but is also used by your web application. If you changed the key in Azure and then in your application, you would have downtime. To avoid this, you can change keys one by one; hence, two keys are provided.

---

5. You are presented with two keys: primary and secondary. Use the primary key in the next command to fetch all available containers in a storage account.

```
azure storage container list -a <storage_account_name>
-k <primary-key>
```

6. Once you have the container name, you can upload or download files from it. The upload file command is as follows.

```
azure storage blob upload -a <storage-account-name>
-k <primary-key><source-file><container-name><blob-name>
Example: azure storage blob upload -a hdistorage
-k mXzSxb.../goS9TSUw== "C:\the_adventures_of_sherlock_
holmes.txt" hdicontainer "books\the_adventures_of_
sherlock_holmes.txt"
```

This example is uploading one of the books from Project Gutenberg, which you can download from www.gutenberg. org/ebooks/1661. Also, notice the addition of the "books/" suffix, which allows you to logically group future books.

7.   Downloading file is somewhat similar and following is
     command for the same.

```
azure storage blob download -a <storage-account-name>
-k <primary-key><container-name><blob-name><destination-file>
Example: azure storage blob download -a hdistorage
-k mXzSxb.../goS9TSUw== hdicontainer "books\the_
adventures_of_sherlock_holmes.txt" "C:\the_adventures
_of_sherlock_holmes.txt"
```

Using Windows PowerShell

Windows PowerShell is becoming popular among DevOps for good reasons. So, let's
start PowerShell and upload a file to Azure Blob Storage. You need to have configured
Windows PowerShell with Azure cmdlets; if you haven't, then you can find the
configuration steps in Chapter 2.

1.   Open Windows PowerShell and type the "Login-AzureRmAccount"
     command to log in. This opens a browser window to enter Azure
     subscription credentials. Once signed in, you can see your
     subscription details in the PowerShell console.

2.   As discussed in the last section, you need a storage account,
     the primary key, and a container to upload files. Execute the
     following command to fetch all available storage accounts.

```
Get-AzureRmStorageAccount | Format-Table
StorageAccountName, ResourceGroupName, Location
```

3.   The default result format of the preceding command is not
     very readable; hence, let's pipe another command: Format-
     Table, which shows data in a tabular format.

4.   Next, get the storage account primary key and save it for later use.

```
$storageAccountKey = (Get-AzureRmStorageAccountKey
-ResourceGroupName<resource-group-name>
-Name <storage-account-name>)[0].Value
```

5.   To upload a file, you need to create a storage context, as
     shown the following. Note that the command uses the storage
     account key variable from the previous step.

```
$context = New-AzureStorageContext -StorageAccountName
<storage-account-name> -StorageAccountKey $storageAccountKey
```

6.   Next, to find the container name, use the following command.

```
Get-AzureStorageContainer -Context $context
```

7. Finally, use the Set-AzureStorageBlobContent command to upload the file.

```
Set-AzureStorageBlobContent -File <file_path> -Container
$containerName -Context $context -Blob <blobname>
Example: Set-AzureStorageBlobContent -File
"C:\the_adventures_of_sherlock_holmes.txt"
-Container $containerName -Context $context -Blob
"books\the_adventures_of_sherlock_holmes.txt"
```

8. Until now, you have only tried uploading a single file. By using a PowerShell script, you can upload all files from a folder. The following command goes through all files and folders inside a specified directory, and then uploads them to Blob storage in the same hierarchy present on the local file system.

```
ls <directory-to-upload> -File -Recurse | Set-
AzureStorageBlobContent -Container $containerName
-Context $context
```

Using Microsoft Azure Storage Explorer

So far, you have seen command-line utilities, but if you want to upload a file quickly without writing all of those commands, then there are several graphical clients as well. One such graphical client is Microsoft Azure Storage Explorer. It is a free tool from Microsoft. It works on Windows, Linux, and Mac OS. It is available at http://storageexplorer.com. It is very easy to configure and use. The following are that steps.

1. Go to http://storageexplorer.com. Download and install Microsoft Azure Storage Explorer.

2. Run Microsoft Azure Storage Explorer. Once you have it open, add the Azure subscription that you want to work with. Change tabs by clicking the icon marked as 1 in Figure 3-3. Then, click the "Add an account..." link.

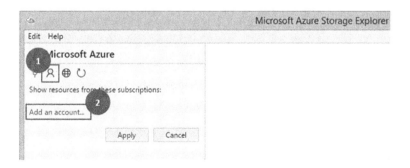

Figure 3-3. Add an Azure subscription to Microsoft Azure Storage Explorer

3. Next is the window where you enter your Windows Azure subscription credentials.

4. Once the credentials are accepted, you should see your subscription(s) listed in the tool, as shown in Figure 3-4. Select a subscription and click the Apply button to manage the storage account(s) available in the subscription.

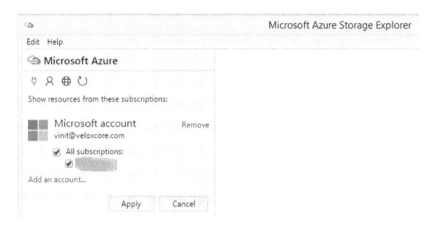

Figure 3-4. *Subscriptions list*

5. Once you select a storage account and a container inside it, you can completely manage the data, including upload a file/folder, download, delete, rename, and so forth, as highlighted in Figure 3-5. Please note that you can manage a Blob container from this tool, but also handle file shares, queues, and tables, as well.

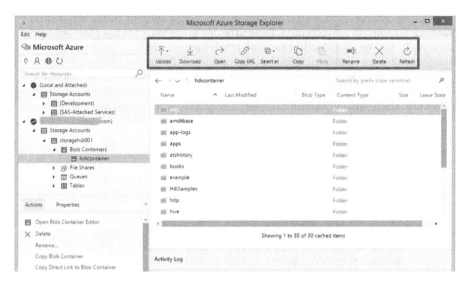

Figure 3-5. *Microsoft Azure Storage Explorer Blob container commands*

Running MapReduce Jobs

MapReduce is a programming framework to process large amounts of data. In a MapReduce process, input data is split into small independent chunks. These chunks are then processed by multiple nodes in parallel in your cluster. A MapReduce job consists of Map ➤ Sort ➤ Shuffle ➤ Reduce phases. As the name suggests, a MapReduce job consists of two functions.

- **Mapper**: A mapper is a function that takes input chunks and outputs a tuple (key-value pairs) by analyzing data, which usually is a filtering and sorting operation. It reads key-value pairs and outputs zero or more key-value pairs.

    ```
    Map(in_key, in_value) -> (inter_key, inter_value) list
    ```

    In the preceding code, the input key may be completely ignored. For example, a standard pattern is to read a file line by line and process it. In such cases, a byte offset of line is key but irrelevant from a processing perspective. On the other hand, the output key cannot be ignored. A mapper should output as key-value pairs only.

- **Reducer**: A reducer consumes tuples emitted by the mapper and performs a summary operation on it. Typically, all the intermediate key-values are combined together into a list. There can be one or multiple reducers. All values associated with a particular intermediate key are guaranteed to go to the same reducer.

Since the reducer executes after the mapper, it is quite possible that some mappers take more time to complete than others do. This can be due to hardware failure or low-powered machines. This can be a bottleneck in the whole process, because the reducer cannot start before all mappers finish execution. To mitigate this problem, Hadoop uses speculative execution. If a mapper is running slower than others are, a new instance is started on another machine, doing the same job on the same data. Results can come from either of the mappers, whichever finishes first. The results are taken from it, and other running mapper processes are killed.

To understand how MapReduce works in conjunction, look at Figure 3-6.

1. First, input data is loaded and divided based on the key-value pairs on which mapping is performed.

2. The output of this process is the intermediate key-value pairs (i.e., List (K2, V2)).

3. Afterward, this list is given to the reducer and all similar keys are processed at the same reducer (i.e., K2, List (V2)). Finally, all the outputs of shuffling are combined together to form a final list of key-value pairs (i.e., List (K3, V3)).

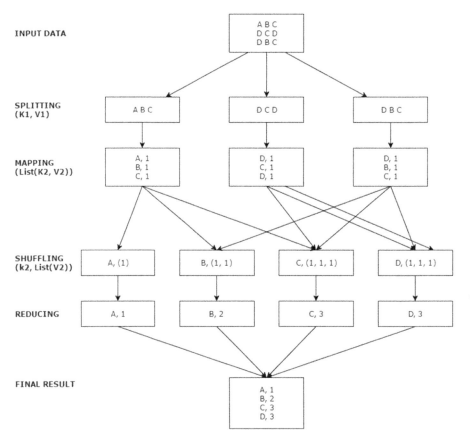

Figure 3-6. *MapReduce word-count process*

■ **Note** Optionally, there is a combiner, which performs local aggregation on the results of the mapper. This reduces the amount of data that needs to be transferred to reducers.

MapReduce is based on a master-and-slave architecture, where JobTracker is the master and TaskTrackers are slaves. When a MapReduce job is submitted, JobTracker, which is running on a master node, does the scheduling, monitoring, and tracking of the execution. JobTracker submits jobs to individual TaskTrackers. Failing tasks are re-executed until a limit, and then the task is submitted to different TaskTrackers. JobTracker also communicates with the name node to understand where the data is present, and then creates an execution plan accordingly. On the other hand, TaskTracker's job is to accept tasks and execute them on slave nodes. TaskTracker can only execute a defined number of tasks in parallel based on slots available (Map/Reduce slots). Each task in TaskTracker is executed in their own Java virtual machine (JVM) process. Due to this separation, if one task goes down, it won't affect other tasks on the same node. Executing tasks are monitored by TaskTracker. Output as well as exit codes are captured here, which are then transferred back to JobTracker and eventually to the job client.

MapReduce jobs are mainly written in JVM-based languages. Hadoop streaming provides a way to write MapReduce jobs in C# or Python. If you already have a jar file that has MapReduce-based code in it, then you can submit it in multiple ways on HDInsight: PowerShell, SSH, cURL, or .NET SDK. PowerShell and .NET SDK are covered in coming sections.

Using PowerShell

Windows PowerShell provides cmdlets that allow you to submit MapReduce jobs to HDInsight from a Windows machine. The following commands will help you submit jobs remotely.

- `Login-AzureRmAccount`: Allows you to authenticate your Azure subscription within PowerShell.

- `New-AzureRmHDInsightMapReduceJobDefinition`: Creates a new MapReduce job object based on the information supplied for execution.

- `Start-AzureRmHDInsightJob`: Starts a defined HDInsight job on a specified cluster by submitting the job to a cluster and returning the job object for future job-status tracking.

- `Wait-AzureRmHDInsightJob`: Waits for the job to finish (success or failure) or until timeout. Accepts a job object.

- `Get-AzureRmHDInsightJobOutput`: Gets the log output for a job from the storage account associated with a specified cluster.

The following script uses the preceding commands and executes a MapReduce job. It uses the word-count example provided in your HDInsight cluster. As input data, it uses the file that you uploaded in the previous section, the_adventures_of_sherlock_holmes.txt. Save the following script as MapReduceWordCount.ps1 in your local hard drive, and then execute it from PowerShell. Also, don't forget to replace the cluster information.

```
# Login to your Azure subscription
$sub = Get-AzureRmSubscription -ErrorAction SilentlyContinue
if(-not($sub))
{
    Login-AzureRmAccount
}

# Set cluster name
$clusterName = "hdi"

# Get HTTP Credential for cluster
$creds = Get-Credential

# Get rest of the cluster details to submit job
$clusterInfo = Get-AzureRmHDInsightCluster -ClusterName $clusterName
$resourceGroup = $clusterInfo.ResourceGroup
$storageAccountName=$clusterInfo.DefaultStorageAccount.split('.')[0]
$container=$clusterInfo.DefaultStorageContainer
$storageAccountKey=(Get-AzureRmStorageAccountKey -Name $storageAccountName
-ResourceGroupName $resourceGroup)[0].Value

# Get the storage context
$context = New-AzureStorageContext -StorageAccountName $storageAccountName
-StorageAccountKey $storageAccountKey

# New job definition
$wordCountJobDefinition = New-AzureRmHDInsightMapReduceJobDefinition
-JarFile "wasbs:///example/jars/hadoop-mapreduce-examples.jar" -ClassName
"wordcount" -Arguments "wasbs:///books/the_adventures_of_sherlock_holmes.
txt", "wasbs:///books/JobOutput/"

# Start the job and start the object
$wordCountJob = Start-AzureRmHDInsightJob -ClusterName $clusterName
-JobDefinition $wordCountJobDefinition -HttpCredential $creds

Write-Host "Wait for the job to complete..." -ForegroundColor Green
Wait-AzureRmHDInsightJob -ClusterName $clusterName -JobId $wordCountJob.
JobId -HttpCredential $creds
```

When prompted, enter the cluster login username and password that was given when the cluster was created. When the job finishes, you should see output similar to what's shown in Figure 3-7.

```
Cluster         : hdi
HttpEndpoint    : hdi.azurehdinsight.net
State           : SUCCEEDED
JobId           : job_1479552910485_0005
ParentId        :
PercentComplete :
ExitValue       : 255
User            : admin
Callback        :
Completed       : done
StatusFolder    : 2016-11-19T11-24-16-320598cc-ef37-4205-be31-2f3af1a5dc9e
```

Figure 3-7. MapReduce job execution output

After the script executes successfully, you should get the results of the MapReduce job at books/JobOutput/part-r-00000 in the storage account. If your job produces multiple files, then they will be in the same location and the suffix number will be incremented accordingly. To download output, you can use either a graphical client or PowerShell. Use the following PowerShell command to download the result (make sure that you are in the same PowerShell window to make use of the context and container variable).The same folder also contains another file, named _SUCCESS. It signifies that the job has completed successfully.

```
Get-AzureStorageBlobContent -Blob 'books/JobOutput/part-r-00000' -Container
$container -Destination output.txt -Context $context
```

Using .NET SDK

Microsoft provides .NET SDK to submit jobs from within your code. This is important because you can embed a job submission in your application without having to rely on other components. Submitting a job from .NET SDK is as equally simple as other methods. Let's create a simple console application to submit a word-count method.

1. Create a C# console application in Visual Studio (use Visual Studio 2012 or higher) targeting the .NET Framework 4.5 or higher.

2. To get SDK bits, install a Nuget package for HDInsight using the Install-Package Microsoft.Azure.Management. HDInsight.Job command. This installs its dependencies as well. The final package.config will look like Figure 3-8.

```xml
<?xml version="1.0" encoding="utf-8"?>
<packages>
  <package id="Hyak.Common" version="1.0.2" targetFramework="net452" />
  <package id="Microsoft.Azure.Common" version="2.1.0" targetFramework="net452" />
  <package id="Microsoft.Azure.Common.Dependencies" version="1.0.0" targetFramework="net452" />
  <package id="Microsoft.Azure.KeyVault.Core" version="1.0.0" targetFramework="net452" />
  <package id="Microsoft.Azure.Management.HDInsight.Job" version="2.0.4" targetFramework="net452" />
  <package id="Microsoft.Bcl" version="1.1.9" targetFramework="net452" />
  <package id="Microsoft.Bcl.Async" version="1.0.168" targetFramework="net452" />
  <package id="Microsoft.Bcl.Build" version="1.0.14" targetFramework="net452" />
  <package id="Microsoft.Data.Edm" version="5.6.4" targetFramework="net452" />
  <package id="Microsoft.Data.OData" version="5.6.4" targetFramework="net452" />
  <package id="Microsoft.Data.Services.Client" version="5.6.4" targetFramework="net452" />
  <package id="Microsoft.Net.Http" version="2.2.22" targetFramework="net452" />
  <package id="Newtonsoft.Json" version="6.0.8" targetFramework="net452" />
  <package id="System.Spatial" version="5.6.4" targetFramework="net452" />
  <package id="WindowsAzure.Storage" version="6.0.0" targetFramework="net452" />
</packages>
```

Figure 3-8. *HDInsight .NET SDK dependencies*

HDInsightJobManagementClient is the main class that facilitates communicating with the HDInsight service. It comes from the Microsoft.Azure.Management.HDInsight. Job namespace. It requires an object of cluster credentials and a cluster Uri to instantiate.

Important methods from the perspective of submitting MapReduce jobs are SubmitMapReduceJob, WaitForJobCompletion and GetJob in HDInsightJobManagementClient.

The following is the complete routine to submit a word-count MapReduce job (on the data uploaded earlier in this chapter) on an existing cluster.

```
private static void SubmitJob()
{
    // Cluster credentials
    Console.WriteLine("Enter cluster http credentils");
    Console.Write("Username: ");
    var clusterUsername = Console.ReadLine();
    Console.Write("Password: ");
    var clusterPassword = GetMaskedPassword();

    // Cluster name
    Console.Write("Enter cluster name: ");
    var clusterName = Console.ReadLine();
    var clusterUri = $"{clusterName}.azurehdinsight.net";

    var clusterCredentials = new BasicAuthenticationCloudCredentials()
    {
        Username = clusterUsername,
        Password = clusterPassword
    };
    HDInsightJobManagementClient jobManagementClient =
                new HDInsightJobManagementClient(clusterUri, clusterCredentials);
```

```
    // Prepare parameters, input, output, jar file and class to use
    var paras = new MapReduceJobSubmissionParameters
    {
        JarFile = @"/example/jars/hadoop-mapreduce-examples.jar",
        JarClass = "wordcount",
        Arguments = new List<string>() {
{ "wasbs:///books/the_adventures_of_sherlock_holmes.txt" },
{ "wasbs:///books/JobOutput" } }
    };

    // Submit job
    Console.WriteLine("Submitting MapReduce job.");
    JobSubmissionResponse jobResponse =

jobManagementClient.JobManagement.SubmitMapReduceJob(paras);
    var jobId = jobResponse.JobSubmissionJsonResponse.Id;
    Console.WriteLine("Response status code is " + jobResponse.StatusCode);
    Console.WriteLine($"Waiting for the job: {jobId} to completion ...");

    // Wait for job to complete
    JobGetResponse waitResponse = jobManagementClient.JobManagement.
    WaitForJobCompletion(jobId);

    // Show job detail
    var jobDetail = waitResponse.JobDetail;
    Console.WriteLine($"{Environment.NewLine}Job Completed.");
    Console.WriteLine($"Status              : {jobDetail.Status}");
    Console.WriteLine($"PercentCompelete    : {jobDetail.PercentComplete}");
    Console.WriteLine($"ExistValue          : {jobDetail.ExitValue}");
    Console.WriteLine($"User                : {jobDetail.User}");
    Console.WriteLine($"Callback            : {jobDetail.Callback}");
    Console.WriteLine($"Completed           : {jobDetail.Completed}");

    // Get job output
    var storageAccess = new AzureStorageAccess("{StorageAccountName}",
        "{StorageAccountKey}", "{StorageContainer}");
    var output = storageAccess.GetFileContent("books/JobOutput/part-r-00000");

    Console.Write("Enter output file path (with file name, i.e. C:\\Output.txt): ");
    var outputFile = Console.ReadLine();
    using (var fileStream = File.Create(outputFile))
    {
        output.Seek(0, SeekOrigin.Begin);
        output.CopyTo(fileStream);
    }
    // Open output file
    System.Diagnostics.Process.Start(outputFile);
}
```

As you can see in the preceding code, it is pretty straightforward to submit a job. Also, the SubmitJob method uses the GetMaskedPassword method to allow the user to type the password on the console without exposing it. The GetMaskedPassword method's code is shown in the following.

```
public static string GetMaskedPassword()
{
    var pwd = new StringBuilder();
    while (true)
    {
        ConsoleKeyInfo i = Console.ReadKey(true);
        if (i.Key == ConsoleKey.Enter)
        {
            Console.WriteLine();
            break;
        }
        else
        {
            pwd.Append(i.KeyChar);
            Console.Write("*");
        }
    }
    return pwd.ToString();
}
```

Finally, the SubmitJob method uses Azure storage to fetch the output file from the output folder defined in the job arguments; in this case, /books/JobOutput.

To run all of these, you just need to call the SubmitJob method from the Main method of your console application.

Hadoop Streaming

So far, you have seen how to submit MapReduce jobs. And you are using a jar file to do the actual work. So, if you want to write your own MapReduce code, then you have to use JVM-based languages. The Hadoop streaming API allows you to use any language (i.e., a language that can read/write to standard input/output). The Hadoop streaming API works with many languages including C#, Ruby, Python, Perl. In this section, you use C# to build your own MapReduce job and submit the same to an HDInsight cluster for processing. But first, let's look at how streaming works.

Hadoop streaming is a utility provided by Hadoop distribution. This utility allows you to provide two executables: mapper and reducer. A mapper gets the input file, line by line, as it is written to stdin. A mapper should output key-value pairs separated by tabs. If it is a word-count program, then it should write each work as a key and "1" as a value, separated by a tab. This output by word is then read by the reducer program and acted upon. If the reducer is a word-count program, then it will group the same words, count their instances, sort them, and then write back to stdout to generate output. Again, reducer also generates key-value pair (i.e., word/count pair) separated by tabs.

One limitation of Hadoop streaming is that it works with stdin/stdout; hence, it can only work with strings or UTF8-encoded bytes. Operations on it should be represented as strings only.

Streaming Mapper and Reducer

To create the mapper and reducer in C#, create two separate console applications. You will re-create the word-count MapReduce job executed earlier. The mapper code should read from stdin and output each word. The mapper code should look like the following snippet.

```
using System;
using System.IO;
using System.Linq;

namespace mapper
{
    class Program
    {
        static void Main(string[] args)
        {
            if (args.Length > 0)
            {
                Console.SetIn(new StreamReader(args[0]));
            }

            string line;

            // Loop through each line and output words
            while ((line = Console.ReadLine()) != null)
            {
                line.Split(new char[] { ' ' })
                    .ToList().ForEach(o => Console.WriteLine($"{o}\t1"));
            }
        }
    }
}
```

This is a very simple mapper, which just loops through all the lines and then outputs a word with value 1. Next, let's look at the reducer code.

```
using System;
using System.Collections.Generic;
using System.IO;
using System.Linq;
```

```csharp
namespace reducer
{
    class Program
    {
        static void Main(string[] args)
        {
            string line;

            if (args.Length > 0)
            {
                Console.SetIn(new StreamReader(args[0]));
            }

            List<Tuple<string, int>> mapOutput = new List<Tuple<string, int>>();
            string[] splitResult;

            // Loop until all lines are processed
            while ((line = Console.ReadLine()) != null)
            {
                splitResult = line.Split('\t');
                mapOutput.Add(new Tuple<string, int>(
                    splitResult[0], Convert.ToInt32(splitResult[1])));
            }

            // Generate output
            mapOutput
                .OrderBy(o => o.Item1)      // Sort result
                .GroupBy(o => o.Item1)       // Group by word to calculate count
                .ToList().ForEach(item => // Loop through all words and
                                          //          output them
                {
                    Console.WriteLine($"{item.Key}\t{item.Count()}");
                });
        }
    }
}
```

The preceding code first loops through all the lines coming from the mapper, and then sorts them and counts the frequency of each word in the entire list. Then, it outputs the word and its frequency, separated by a tab, to stdout.

To execute C# mapper and reducer programs, you can use any of the available job submission methods discussed earlier in this chapter. To use PowerShell, you can replace the New-AzureRmHDInsightMapReduceJobDefinition line in the script with the following command.

```
$wordCountJobDefinition = New-AzureRmHDInsightStreamingMapReduceJobDefinition
-Files "/example/streaming/mapper.exe","/example/streaming/reducer.exe"
-Mapper "mapper.exe" -Reducer "reducer.exe" -InputPath "/books/the_
adventures_of_sherlock_holmes.txt" -OutputPath "/books/StreamingOutput/"
```

Notice that this snippet uses mapper.exe and reducer.exe, which you already uploaded to Azure Blob storage under example/streaming.

To execute a streaming job using .NET, the SubmitMapReduceStreamingJob method is available in Microsoft.Azure.Management.HDInsight.Job.HDInsightJob ManagementClient's JobManagement object.

Serialization with Avro Library

Apache Avro is a language-neutral data serialization system mostly used in Hadoop environments. It provides a compact binary data interchange format for serialization. *Language-neutral* means data serialized in one language can be read in another one. Currently C, C++, C#, Java, PHP, Python, and Ruby are supported. Serialization output with Avro is both compact and extensible. It is compact in terms of the number of bytes it takes to store in plain text. *Extensible* refers to the schema that can evolve without negatively impacting existing serialized data.

An Avro serialized representation of an object has two parts: schema and value. The schema part holds the description of the data model in a language-independent JSON format. It is presented side-by-side with a binary representation of data. Unlike JSON, where each record holds its schema within it, Avro keeps it once per file, avoiding the overhead of extra characters, repeating for each record, making serialization a fast and small footprint.

Data Serialization

Avro data is always serialized with its schema. Files that store Avro data should always include the schema for that data in the same file. Since schema is stored with data, even if the schema changes, existing data can be read using a schema stored in the Avro data file.

Avro has two types of serialization encoding: binary and JSON. Binary encoding is faster and produces a smaller footprint, but it is not human readable. On the other hand, JSON encoding is used for debugging and web-based applications.

Binary Encoding

Binary encoding is widely used in Hadoop environments due to its unique benefits. Primitive types are encoded in binary as follows.

- null is written as 0 bytes.

- boolean is written as a single byte whose value is either 0 (false) or 1 (true).

- int and long values are written using variable-length zigzag coding. Zigzag encoding maps signs integers to unsigned integers so that numbers with a small absolute value (for instance, -1) have a small variant encoded value too. So, -1 is encoded as 1, 1 is encoded as 2, -2 is encoded as 3, and so on. In other words, each n value is encoded using

  (n << 1) ^ (n >> 31)

Note that the second shift is an arithmetic shift (`https://en.wikipedia.org/wiki/Arithmetic_shift`). So, the result of a shift is either a number that is all 0 bits (if n is positive) or all 1 bit (if n is negative). A left arithmetic shift moves binary numbers left by 1 and fills a vacant bit with 0 (zero). A right arithmetic shift moves binary numbers right by 1 and fills vacant a bit with the original bit in the leftmost position.

For example, let's say that n is 5:

```
(5 << 1) ^ (5 >> 31)
```

If you represent it in binary, then it will look like following,

```
(00000101 << 1) ^ (00000101 >> 31)
(00001010) ^ (00000000)
```

00001010 converted to decimal results in 10.

Similarly, if you try n = –5

```
(-5 << 1) ^ (-5 >> 31)
(11111010 << 1) ^ (11111010 >> 31)
(11110100) ^ (11111111)
```

000001011 in decimal is 11.

- `float` is written as 4 bytes.

- `double` is written as 8 bytes.

- bytes are encoded as a long count value followed by that many bytes of data.

- `string` is encoded as a long count value followed by that many bytes of UTF-8 encoded character data. For example, a `"bar"` string would be encoded as a value 3 (encoded as hex 06 as per integer conversion) followed by 3 UTF-8 encoded char `'b'`, `'a'`, and `'r'`, resulting in 62, 61, 72; hence, a complete encoding of a `"bar"` string would be a 06 62 61 72 block.

- `enum` is encoded as int, representing the zero-based position of the symbol in the schema. For example, consider this enum:

  ```
  {"type": "enum", "name": "bar", "symbols": ["A", "B", "C", "D"]}
  ```

  This would be an int between 0 and 3, with A indicating 0 and D indicating 3.

- arrays are encoded as a series of blocks, which consist of a long count value followed by that many array items. For example, array schema

```
{"type": "array", "items": "long"}
```

An array with value 6 and 14 could be encoded as a long value 2 (hex 04) followed by a long value 6 and 14 (hex value 0C and 1C) terminated by 0 (hex 00).

```
04 0C 1C 00
```

- union is typically used to represent either of the types. Let's say that you want to present a nullable string, then a union schema would be ["null", "string"]. Encoding first writes a long value indicating the zero-based position within the union of the schema, and followed by the value encoded per the indicated schema within the union. For example, schema ["null", "string"].

  - null as 0 (the index of "null" in the union)

    ```
    00
    ```

  - The string "a" as one (hex 02) followed by UTF8 of char "a" (encoded as 61)

    ```
    02 02 61
    ```

- Consider the following JSON object.

  ```
  {"firstname": "Scott","age": 25,"contact": ["email", "phoneno"]}
  ```

The Avro JSON schema of the same object can be written as follows.

```
{
    "type": "record".
    "name": "Person",
    "fields": [
            {"name": "firstname",          "type": "string"},
            {"name": "age",                "type": ["null", "long"]},
            {"name": "contact",            "type": {"type": "array",
                                                    "items": "string"}}

            ]
}
```

Binary encoding of the preceding record would look like Figure 3-9.

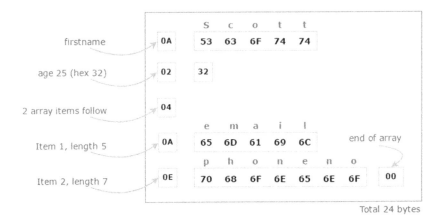

Figure 3-9. *Sample Avro object encoding*

As you can see from Figure 3-9, strings are nothing but bytes. There is no indication whether it is a string or something else. To make sense out of this binary data, read it based on the schema provided. The schema tells the reader what to expect next.

As far as schema evolution is concerned, there can be two schemas: the writer's schema and the reader's schema. That means there can be two different schemas at parsing time. Hence, if there is a type mismatch in the reader's and writer's schemas, it will use schema resolution rules (more information at http://bit.ly/2g61dp4).

JSON Encoding

JSON encoding is the usual encoding that you do in any language. An object value is represented in the default value field, unless it is a union. In a union type, null is encoded as JSON null; otherwise, it is encoded as a JSON object with the type's name and value pair.

For example, the union schema ["null", "string", "bar"] where bar is a record name would be encoded as follows.

- null as null

- String "a" as {"string": "a"} and

- Bar instance as {"bar": {…}}, where {…} instances the JSON encoding of bar instance.

Using Microsoft Avro Library

Avro provides a convenient way to represent a complex data structure in a Hadoop MapReduce job. Avro is specifically designed to handle distributed MapReduce programming models. It does this by providing ability to split file in chunks; that is, you can seek any point in a file and start reading from a particular block. The .NET library for Avro supports two types of serialization: based on reflection and based on generic record.

In the reflection method, a JSON schema for the types is automatically built from the data contract attribute of .NET types. On the other hand, in generic record-based serialization, a JSON schema is specified manually for serialization and deserialization. Here, I discuss generic record-based serialization.

Let's look at the following JSON schema.

```
{
"type":"record",
"name":"movie",
"fields":
        [
                {"name": "title", "type": "string"},
                {"name": "director", "type": "string"},
                {"name": "duration", "type": "int"},
                {"name": "gross", "type": "double"},
                {"name": "genres", "type":
                        {"type": "array", "items": "string"}},
        ]
}
```

Let's try to serialize data from a .csv file to the preceding JSON schema using Avro. This is a typical scenario in Hadoop, in which you take raw data (CSV, plain text, or JSON) and use some serialization system to convert and store them for processing by a MapReduce job. The output of a serialization operation is another file, which you can upload to a storage container for a MapReduce job to process it. You will use open movie data published by Kaggle (www. kaggle.com). It can be downloaded from http://bit.ly/2c6YmXj.

To demonstrate usage of Microsoft Avro Library, let's use Visual Studio 2015 and create a C# console app targeting .NET Framework 4.5. Once you have the console application ready, add a Nuget package using the following command.

```
Install-package Microsoft.Hadoop.Avro
```

You will create two methods: SerializeMovieData and DeserializeMovieData, which will perform serialization and deserialization, respectively. Both of these methods use common schema defined in JSON and stored in a string constant. They also use a couple of constants to locate input and output file paths. Schema and path constants are shown in the following.

```
const string sourceFilePath = @"C:\OpenData\movie_metadata.csv";
const string outputFilePath = @"C:\OpenData\movie_metadata.avro";

const string schema = @"{
                    ""type"":""record"",
                    ""name"":""movie"",
                    ""fields"":
                        [
                            {""name"": ""title"", ""type"": ""string""},
```

```
                         {""name"": ""director"", ""type"": ""string""},
                         {""name"": ""duration"", ""type"": ""int""},
                         {""name"": ""gross"", ""type"": ""double""},
                         {""name"": ""genres"", ""type"":
                              {""type"": ""array"", ""items"": ""string""}},
                    ]
               }";
```

Let's first look at a serialize method to read a .csv file, and convert it to an AvroRecord list. As this demonstrates generic record serialization, you will use dynamic to define properties, because you don't want to reply on a predefined type. After reading the data, you will use CreateGenericWriter to write an object to Avro blocks.

```
public void SerializeMovieData()
{
    //Create a generic serializer based on the schema
    var serializer = AvroSerializer.CreateGeneric(schema);

    var avroRecords = new List<AvroRecord>();

    // Read all csv data and create AvroRecord list
    File.ReadAllLines(sourceFilePath)
        .Select(line => line.Split(',')).Skip(1)
        .Where(o => !string.IsNullOrEmpty(o[3]) && !string.IsNullOrEmpty(o[8]))
        .ToList().ForEach(o =>
        {
            dynamic avroRecord = new AvroRecord(serializer.WriterSchema);
            avroRecord.title = o[11];
            avroRecord.director = o[1];
            avroRecord.duration = Convert.ToInt32(o[3]);
            avroRecord.gross = Convert.ToDouble(o[8]);
            avroRecord.genres = o[9].Split('|').ToArray<string>();
            avroRecords.Add(avroRecord);
        });

    //Create a memory stream buffer
    using (var stream = new MemoryStream())
    {
        using (var writer = AvroContainer.CreateGenericWriter(schema,
        stream, Codec.Null))
        {
            using (var streamWriter = new SequentialWriter<object>(writer, 24))
            {
                // Serialize records
                avroRecords.ForEach(streamWriter.Write);
            }
        }
    }
```

```
        // Remove file if not available
        if (File.Exists(outputFilePath))
            File.Delete(outputFilePath);

        using (FileStream fs = File.Create(outputFilePath))
        {
            stream.Seek(0, SeekOrigin.Begin);
            stream.CopyTo(fs);
        }
    }
}
```

After you execute the SerializeMovieData method, you should have a new file in the output directory. And if you open this file in Notepad, you can see the schema in the top few lines of the file.

Now, the output file can be deserialized in any of the languages supported by Avro. The following is another method, DeserializeMovieData, which shows how to read an Avro file using generic record serialization.

```
public void DeserializeMovieData()
{
    //Reading and deserializing the data.
    //Create a memory stream buffer.
    using (var buffer = new MemoryStream())
    {
        Console.WriteLine("Reading data from file...");
        using (FileStream fs = File.Open(outputFilePath, FileMode.Open))
        {
            fs.CopyTo(buffer);
        }

        //Prepare the stream for deserializing the data
        buffer.Seek(0, SeekOrigin.Begin);

        // Create a SequentialReader instance, which will deserialize all
        serialized objects.
        // It allows iterating over the deserialized objects
        // because it implements the IEnumerable<T> interface.
        using (var reader = AvroContainer.CreateGenericReader(schema,
        buffer, true, new CodecFactory()))
        {
            using (var streamReader = new SequentialReader<object>(reader))
            {
                var results = streamReader.Objects;
                foreach (dynamic pair in results.Take(10))
                {
                    Console.WriteLine($"{pair.title},{pair.director}" +
```

```
                        $",{pair.duration},{pair.gross},{string.Join("|",
                        pair.genres)}");
                }
            }
        }
    }
}
```

If you run both the serialize and deserialize methods in the same console app, you would see ten movie titles and their information in the console. This information is read from the Avro serialized file.

Summary

Azure Blob storage is a general-purpose and robust storage option. Working with Azure Blob storage is much easier than working with the HDFS file system. Graphical clients make it even more convenient for beginners to start using it without remembering all the commands.

Apache Avro, or more general encoding, plays an important part in the Hadoop environment. From saving disk space, increasing the performance of queries, to improving data saving and retrieval speed, data encoding is crucial. C# developers can easily work with Avro encoding using Microsoft Avro Library.

MapReduce provides a processing framework to work with large amount of data in a distributed manner. Creating your own MapReduce job requires you to learn a JVM-based language. For C# developers, Hadoop streaming is one of the options for writing Map and Reduce code. On the other hand, submitting a MapReduce job on HDInsight can be done in C# directly, even for MapReduce code written in Java language.

Next, let's look at how to work with data without writing tedious MapReduce code.

CHAPTER 4

▪ ▪ ▪

Querying Data with Hive

Hive is probably the most used tool in the Hadoop ecosystem. To work with Hadoop data, you need to write MapReduce jobs that are not convenient for ad hoc queries. Hive comes to the rescue by providing a SQL-like query language, which internally transforms the query to MapReduce jobs. In HDInsight, Hive sits on top of Azure Blob storage data and provides interactive queries to work with data. Hive can work with structured and semi-structured data. Hive resides on top of a YARN layer and makes use of all the resource negotiations that YARN does. Internally, it uses MapReduce, Tez, or the Spark execution engine.

MapReduce is very good at processing large volumes of data. While it sounds great, the only problem is that you have to learn to write MapReduce programs, which can be very low level, so SQL programmers don't like it. Also, there are business analysts and BI folks who are comfortable in writing SQL queries. With Hive, all users with knowledge of SQL can query data from HDInsight. Hive can run SQL queries but it is not a relational database; actually, it doesn't even store data by itself. Hive uses data stored on HDInsight and maps a schema on top of it. This schema is stored in an external SQL database or in HCatalog. The scripting language of Hive is called HiveQL.

Hive uses a schema-on-read approach, which means that you can create a table any time, whether there is existing data or no data at all. A schema is applied on data when you try to query it out. This is completely different from a traditional RDBMS, where you first define that schema and then insert data. Schema-on-read implies that you can dump your data in whatever format you want and worry about it later when you try to query it. You can even have files with different formats and work with them as well.

Hive Essentials

Hive is effective with structured and semi-structured data (i.e., web log or click-stream processing). Let's understand Hive better by diving into the code. First, let's get data for processing. Download the movie data from https://www.kaggle.com/deepmatrix/imdb-5000-movie-dataset. You are using this data throughout this chapter. It is mostly structured data, except a few rows have a different column format. Also, keep only the columns with director_name, duration, gross, movie_title, language, country, content_rating, title_year, and IMDB_score. Delete the rest of the columns. Upload it to your HDInsight cluster under the movie_data folder by using any of the methods discussed in previous chapters. After you have your data in HDInsight, open an SSH session (if you have a Windows cluster, then you can also go to the dashboard and open the Hive View

© Vinit Yadav 2017
V. Yadav, *Processing Big Data with Azure HDInsight*, DOI 10.1007/978-1-4842-2869-2_4

to execute Hive queries). Once you are in your cluster through SSH, write Hive to get into the Hive console. Inside the Hive console, write the following query to create a table that is stored in the movie_data folder in your Azure Blob storage.

```
CREATE EXTERNAL TABLE movies
(
director STRING, duration INT, gross DECIMAL,
genres STRING, title STRING, language STRING,
country STRING, ratings STRING, budget DECIMAL,
release_year INT, score FLOAT
)
ROW FORMAT DELIMITED
FIELDS TERMINATED BY ','
LINES TERMINATED BY '\n'
STORED AS TEXTFILE LOCATION 'wasb:///movie_data/';
```

When you execute the preceding query, you get a new table in our default database. The default database is already present in your HDInsight cluster and has a table named hivesampletable. To view the number of databases that you have, execute the show databases; command; similar to how you show tables inside the database—execute shows tables;. The following is the result of our current HDInsight cluster.

```
hive> show tables;
OK
hivesampletable
movies
Time taken: 0.145 seconds, Fetched: 2 row(s)
```

You can see that our newly created movies table is also present in the list. Next, let's query data from the newly created table. Notice that it utilizes the CSV file that you uploaded earlier to Azure Blob storage in the movie_data folder. Let's first fetch ten rows and a few columns. The query and results are shown in Figure 4-1.

```
hive> SELECT director, duration, gross, release_year, ratings, title FROM movies LIMIT 10;
OK
director_name      NULL     NULL     NULL      content_rating  movie_title
James Cameron      178      760505847          2009      PG-13    Avatar
Gore Verbinski     169      309404152          2007      PG-13    Pirates of the Caribbean: At World's End
Sam Mendes         148      200074175          2015      PG-13    Spectre
Christopher Nolan           164      448130642          2012      PG-13    The Dark Knight Rises
Doug Walker        NULL     NULL     NULL                Star Wars: Episode VII - The Force Awakens
Andrew Stanton     132      73058679           2012      PG-13    John Carter
Sam Raimi          156      336530303          2007      PG-13    Spider-Man 3
Nathan Greno       100      200807262          2010      PG       Tangled
Joss Whedon        141      458991599          2015      PG-13    Avengers: Age of Ultron
Time taken: 0.56 seconds, Fetched: 10 row(s)
```

Figure 4-1. *Select query result of movies table*

Notice that the preceding data also shows row headers from the CSV file as one record. Also, there NULL appears in the duration, gross, and release_year columns, because you specified these columns' data types as INT, DECIMAL, and INT, respectively, and the data is not compatible; hence, Hive puts a NULL in place. Another thing to note is row 6, where it says "Doug Walker." Everything else is null except movie_title. If you look into the data file, it doesn't have any data in the columns; hence, Hive put NULL there.

The important thing to note is that it returned results in subseconds because it is not creating MapReduce jobs; instead, it is smart enough to get records from the file. If you want Hive to execute MapReduce jobs, then you have to execute a query that does some calculation. Take a query like SELECT COUNT(*) FROM movies;, for example. This query calculates the total rows using MapReduce; these results are shown in Figure 4-2.

```
hive> set hive.execution.engine=MR;
hive> SELECT COUNT(*) FROM movies;
Query ID = sshuser_20161209123605_e3157489-fd91-489e-b0f0-e3c989231960
Total jobs = 1
Launching Job 1 out of 1
Number of reduce tasks determined at compile time: 1
In order to change the average load for a reducer (in bytes):
  set hive.exec.reducers.bytes.per.reducer=<number>
In order to limit the maximum number of reducers:
  set hive.exec.reducers.max=<number>
In order to set a constant number of reducers:
  set mapreduce.job.reduces=<number>
Starting Job = job_1481280591580_0008, Tracking URL = http://hn1-hdi1.t4wtt5kxoafehhg1i4somrw
kpd.hx.internal.cloudapp.net:8088/proxy/application_1481280591580_0008/
Kill Command = /usr/hdp/2.5.2.1-1/hadoop/bin/hadoop job  -kill job_1481280591580_0008
Hadoop job information for Stage-1: number of mappers: 1; number of reducers: 1
2016-12-09 12:36:25,825 Stage-1 map = 0%,  reduce = 0%
2016-12-09 12:36:35,278 Stage-1 map = 100%,  reduce = 0%, Cumulative CPU 4.56 sec
2016-12-09 12:36:46,327 Stage-1 map = 100%,  reduce = 100%, Cumulative CPU 11.98 sec
MapReduce Total cumulative CPU time: 11 seconds 980 msec
Ended Job = job_1481280591580_0008
MapReduce Jobs Launched:
Stage-Stage-1: Map: 1  Reduce: 1   Cumulative CPU: 11.98 sec   HDFS Read: 0 HDFS Write: 5 SUC
CESS
Total MapReduce CPU Time Spent: 11 seconds 980 msec
OK
5044
Time taken: 42.769 seconds, Fetched: 1 row(s)
hive>
```

Figure 4-2. *Hive MapReduce job*

Please note the time that it took to execute the job, which was around 42 seconds, you revisit the same query when you look at the Apache Tez execution engine later in the chapter. Tez is the default execution engine, for Hive, but you set MapReduce as the execution engine and then executed the preceding query. You are using the MapReduce engine until you learn about the Tez engine later in the chapter.

Now you might be wondering that to calculate only 5044 rows it took 42 seconds. RDBMS could have done this in milliseconds. Hive is designed to work with millions of records in distributed manner. Overhead of submitting MapReduce job to worker nodes and getting result back is quite big in case of small dataset just like in preceding query, hence it took 42 seconds. Later in the chapter, you look at the Tez engine and other improvements, such as LLAP (see http://hortonworks.com/blog/llap-enables-sub-second-sql-hadoop/) happening in Hive makes it possible to get results in subseconds.

Hive Architecture

Hive sits on top of HDFS. To be more precise it works on top of YARN, as you can see in Figure 4-3. Any Hive query submitted to a Hadoop cluster gets compiled and optimized, which is then distributed to the worker nodes so that work can be parallelized. As seen in Figure 4-3, there are a number of ways that you can interact with Hive, which includes ODBC, JDBC (through Thrift server), the Hive web interface, Hive Command Line, and so forth.

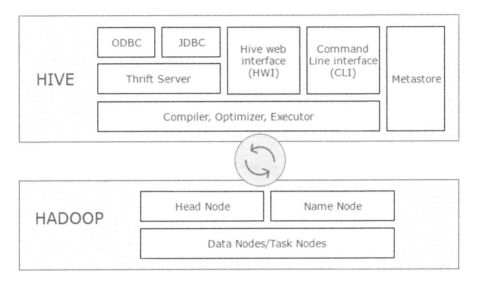

Figure 4-3. *Hive structure in Hadoop cluster and components*

As a user, when you submit a Hive query (which can come from any of the components like Hive Command Line, PowerShell, JDBC/ODBC, Excel or Web UI), then it is actually parsed and Hive prepares an execution plan. To create the execution plan, Hive reads its metadata store (HCatalog or Azure SQL Database), which you've already defined, and then compiles and optimizes it. So essentially, this is the phase where Hive SQL query is converted to a MapReduce program or a Tez program and then executed through a driver program in Hive. Figure 4-4 shows typical query execution steps.

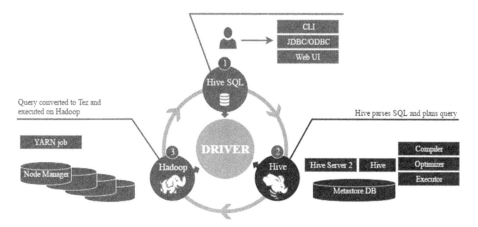

Figure 4-4. *Hive query execution steps*

Figure 4-5 shows the driver and Hive components. The driver sits between all the components. It takes input from the client by any of the available ways that a user can submit a query. This passes to the compiler. Utilizing the metastore generates an execution plan and the actual MapReduce/Tez program. It then submits to Hadoop and returns the results back to the client.

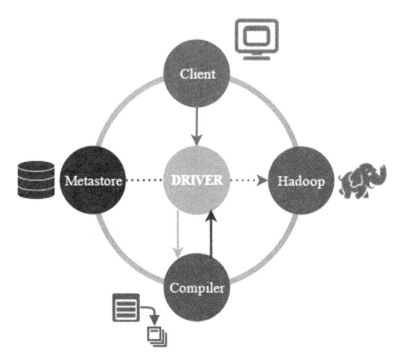

Figure 4-5. *Hive components*

So, that is Hive. The following are typical use cases for Hive:

- Log processing

- Text mining

- Document indexing

- Customer-facing business intelligence (e.g., Google Analytics)

- Predictive modeling/hypothesis testing

Submitting a Hive Query

In HDInsight, you can run a HiveQL query using a variety of tools. Table 4-1 shows all the possible tools and their usages. It includes the cluster type on which they are supported.

Table 4-1. *Hive Tools*

Tool	Cluster Operating System	Client Operating System
Hive View	Linux	Any browser-based
Beeline command (from an SSH session)	Linux	Linux, Unix, Mac OS X, or Windows
Hive command (from an SSH session)	Linux	Linux, Unix, Mac OS X, or Windows
cURL	Linux or Windows	Linux, Unix, Mac OS X, or Windows
Query console	Windows	Any browser-based
HDInsight tools for Visual Studio	Linux or Windows	Windows
Windows PowerShell	Linux or Windows	Windows
Remote Desktop	Windows	Windows

Using Hive View

The Ambari management and monitoring utility provided with Linux-based clusters provides Hive View, through which you can execute Hive queries from any browser. To open the Ambari dashboard, go to `https://CLUSTERNAME.azurehdinsight.net`. In the Azure portal in a HDInsight cluster, look for Ambari Views in the Quick Links, as shown in Figure 4-6.

Figure 4-6. *HDInsight dashboard quick links*

Once you are on the Ambari dashboard, select the set of squares from the Page menu (a button on the left of the page) to list the available views. Select the Hive View, as shown in Figure 4-7.

Figure 4-7. *Ambari Hive View menu item*

Figure 4-8 shows Hive View from where you can submit a Hive query for execution. Also on the left side, you can see all the databases and their tables. Clicking a table shows its columns. With HDInsight cluster, you get the default database. And unless you specify the database name to query, they execute against the default database. Also, for a new HDInsight cluster, you get one sample table, `hivesampletable`. On the right-hand side, you can see other tabs, such as Settings, Visual Explain, Tez, Notifications. Explore these sections to get familiar with the tool.

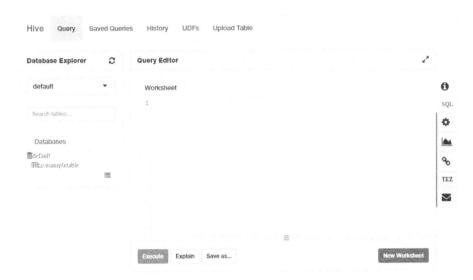

Figure 4-8. *Hive web view*

To see it in action, write a simple query: SELECT * FROM hivesampletable LIMIT 10; . Click the Execute button. Submit the query and get results back.

> ■ **Note**   When you execute a job and it appears to run forever without updating the log or returning results, then try the latest browser.

In Figure 4-8, you see a menu at the top of the screenshot, showing Query as the current tab. Saved Queries stores the queries that you execute again and again. History keeps track of all the queries that you have submitted so far in the current session. UDF is User Defined Functions, which allows you to implement functionality or logic that isn't easily modeled in HiveQL. And last, Upload Table allows you to upload a file from a local machine or from HDFS to create a table based on data (CSV, JSON, or XML).

Using Secure Shell (SSH)

Chapter 2 discussed how to connect to your HDInsight cluster using SSH. Follow the same steps to log in using SSH. Once you are connected to the cluster, use the Hive command to get into the Hive query console. Figure 4-9 shows a SSH connection and the result of the Hive command. Note that many lines were removed for brevity.

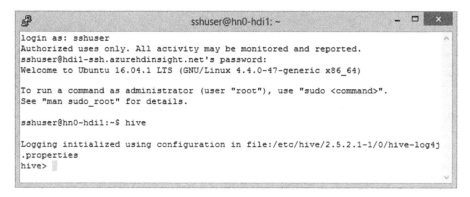

Figure 4-9. SSH using PuTTY on Windows with a Hive query console

Using Visual Studio

So far, you have seen the command-line tools that interact with Hive, but if you want, you can use Visual Studio as well. To get the necessary tools, go to the Azure portal and open your HDInsight cluster. Inside the Quick Start page, you can find a link to download Azure SDK for .NET. Clicking this link downloads the Microsoft Web Platform installer, which gives you the option to download Azure SDK for Visual Studio 2013/15/17. Install the appropriate version as per the Visual Studio version you have on your machine. I'm working with Visual Studio 2015, but you can also choose the 2013 or 2017 versions.

After installing Azure SDK for .NET, open Visual Studio and navigate to Server Explorer. You may need to enter your Azure subscription credentials. After that, you should be able to view your cluster under the HDInsight node. Expand your cluster node. After that, expand the Hive database node. Now you can see all the databases in Hive. If you are viewing a newly created cluster, then it should have a default database. Inside that, you can find the hivesampletable table, as shown in Figure 4-10.

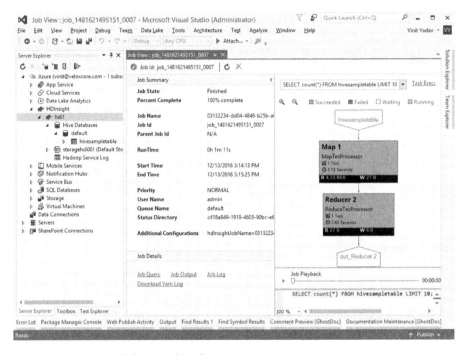

Figure 4-10. *Hive tools for Visual Studio*

To execute a new query, right-click the cluster name and choose Write a Hive Query, which opens the Hive query editor. Note that this is not a simple text editor; you get full intelligence based on your HDInsight cluster. Let's look at this tool in action.

1. Start by opening the Hive query editor and write "SELECT * FROM hivesampletable LIMIT 10;".

2. Click the Submit button available on the top bar. Then select the correct HDInsight cluster from the drop-down menu beside the Submit button.

3. After the job finishes, you see the job results with summaries, execution graphs, and other related details, as shown in Figure 4-10. Execution graph is interactive graph. You can use the Job Playback button to actually see how your job was executed. This is really an interesting feature, because at times this visual representation helps you understand where your query lags and where the bottlenecks are present.

If you are not interested in the graph and all of this information, and you only want to see output from the query, then you can choose to run it in interactive mode. To change modes, there is a drop-down in the top panel inside the Hive query editor. Executing a query in interactive mode only shows you the results of the query.

Using .NET SDK

The HDInsight .NET SDK provides libraries that make it easy to work with Hive. You can submit queries programmatically, allowing you to easily integrate a Hive query in your application flow. Let's try to submit a simple query and get output from it.

1. Create a C# console application in Visual Studio (use Visual Studio 2012 or higher) targeting .NET framework 4.5 or higher.

2. To get SDK bits, install the Nuget package for HDInsight using the `Install-Package Microsoft.Azure.Management.HDInsight.Job` command. This installs its dependencies as well.

3. `HDInsightJobManagementClient` is the main class, which facilitates communicating with the HDInsight service and comes from namespace `Microsoft.Azure.Management.HDInsight.Job`. It requires object of cluster credentials and cluster Uri to instantiate.

4. The important methods from the perspective of submitting a Hive query job are `SubmitHiveJob`, `WaitForJobCompletion`, and `GetJob` in `HDInsightJobManagementClient`.

The following is the complete routine to submit a Hive query to fetch the first 10 records from the `hivesampletable` table.

```
private static void SubmitHiveJob()
{
    // Cluster credentials
    Console.WriteLine("Enter cluster http credentils");
    Console.Write("Username: ");
    var clusterUsername = Console.ReadLine();
    Console.Write("Password: ");
    var clusterPassword = GetMaskedPassword();

    // Cluster name
    Console.Write("Enter cluster name: ");
    var clusterName = Console.ReadLine();
    var clusterUri = $"{clusterName}.azurehdinsight.net";

    var clusterCredentials = new BasicAuthenticationCloudCredentials()
    {
        Username = clusterUsername,
        Password = clusterPassword
    };
    HDInsightJobManagementClient jobManagementClient =
                new HDInsightJobManagementClient(clusterUri,
                clusterCredentials);
```

```
var parameters = new HiveJobSubmissionParameters
{
    Query = "SELECT * FROM hivesampletable LIMIT 10;"
};

System.Console.WriteLine("Submitting the Hive job to the cluster...");
var jobResponse = jobManagementClient.JobManagement.SubmitHiveJob
(parameters);
var jobId = jobResponse.JobSubmissionJsonResponse.Id;
System.Console.WriteLine("Response status code is " + jobResponse
.StatusCode);
System.Console.WriteLine("JobId is " + jobId);

System.Console.WriteLine("Waiting for the job completion ...");

// Wait for job completion
JobGetResponse waitResponse = jobManagementClient.JobManagement
.WaitForJobCompletion(jobId);

// Get job output
var storageAccess = new AzureStorageAccess("{StorageAccount}",
"{StorageAccountKey}", "{StorageAccountContainer");
// fetch stdout output in case of success or stderr output in case of
failure
var output = (waitResponse.JobDetail.ExitValue == 0)
    ? jobManagementClient.JobManagement.GetJobOutput(jobId,
    storageAccess)
    : jobManagementClient.JobManagement.GetJobErrorLogs(jobId,
    storageAccess);

System.Console.WriteLine("Job output is: ");

using (var reader = new StreamReader(output, Encoding.UTF8))
{
    string value = reader.ReadToEnd();
    System.Console.WriteLine(value);
}
}
```

Writing HiveQL

The Hive Query Language (HiveQL) is a SQL-like language for writing Hive jobs. It does not support the complete SQL standard (SQL-92) but the support is improving day by day. Even with the current version of Hive, you get most of the SQL syntax. So, let's look at all the things that are available and how to use HiveQL. I am using Hive version 0.14 (the latest at the time of writing).

Data Types

Hive has a wide range of data types that cover almost all the cases that you can think of. The following list shows all available data types.

- Numeric types

    - TINYINT: 1-byte signed integer

    - SMALLINT: 2-byte signed integer

    - INT/INTEGER: 4-byte signed integer

    - BIGINT: 8-byte signed integer

    - FLOAT: 4-byte single precision floating point number

    - DOUBLE: 8-byte double precision floating point number

    - DECIMAL: Introduced in Hive 0.11.0 with a precision of 38 digits

- Date/time types

    - TIMESTAMP: A traditional UNIX timestamp with optional nanosecond precision.

    - DATE: A particular year/month/day, in the form YYYY-MM-DD.

    - INTERVAL: Intervals of time units— second/minute/day/month/year.

- String types

    - STRING: String literals, expressed with either single quotes (') or double quotes (").

    - VARCHAR: Created with a length specified between 1 and 65355.

    - CHAR: A fixed-length and similar to varchar, up to 255 max. Shorter than the specified length values are padded with spaces.

- Miscellaneous types

    - BOOLEAN

    - BINARY

- Complex types

  - ARRAY<data_type>: A collection of the same type values.

  - MAP<primitive_type, data_type>: A dictionary of key-value pairs.

  - STRUCT<col_name : data_type, ... >: A structs with a different column.

  - UNIONTYPE<data_type, data_Type, ...>: A combination of multiple values of the same/different data types.

Create/Drop/Alter/Use Database

Apache Hive includes a default database named "default". You can also create new databases as and when needed. You would create new database when you want to separate data for different applications. HiveQL provides create, alter, and drop database options. The following is the syntax for the same.

- CREATE: Allows you to create a new database.

  ```
  CREATE (DATABASE|SCHEMA) [IF NOT EXISTS] database_name [COMMENT
  database_comment][LOCATION hdfs_path][WITH DBPROPERTIES
  (property_name=property_value, ...)];
  ```

- IF NOT EXISTS: Statement creates a database only if it doesn't exist already.

- ALTER: A database allows you to change the properties and owner of the database. This is the syntax:

  ```
  ALTER (DATABASE|SCHEMA) database_name SET DBPROPERTIES
  (property_name=property_value, ...);
  ALTER (DATABASE|SCHEMA) database_name SET OWNER [USER|ROLE]
  user_or_role;
  ```

- DROP: Removes a database permanently. The default behavior is RESTRICT, which means if there is table in the database, drop fail. To remove tables as well, use the CASCADE option.

  ```
  DROP (DATABASE|SCHEMA) [IF EXISTS] database_name
  [RESTRICT|CASCADE];
  ```

- Use database sets the current database as default database for subsequent queries, where you don't specify database to use to execute query upon.

  ```
  USE database_name;
  ```

■ **Note**   In Hive, both DATABASE and SCHEMA refer to the same thing. That means
CREATE DATABASE movies and CREATE SCHEMA movies create a new database named
movies.

The Hive Table

Hive provides two types of tables: internal and external. Type of table you want to create
depends on data retention mechanism you want for the table when the table is deleted.
The following is the table-creation query syntax. Examples are provided in upcoming
sections.

```
CREATE [TEMPORARY] [EXTERNAL] TABLE [IF NOT EXISTS] [db_name.]table_
  [(col_name data_type [COMMENT col_comment], ...)]
  [COMMENT table_comment]
  [PARTITIONED BY (col_name data_type [COMMENT col_comment], ...)]
  [CLUSTERED BY (col_name, col_name, ...) [SORTED BY (col_name [ASC|DESC],
...)]
INTO num_buckets BUCKETS]
[STORED AS DIRECTORIES]
  [
  [ROW FORMAT row_format]
  [STORED AS file_format]
  ]
  [LOCATION path]
  [TBLPROPERTIES (property_name=property_value, ...)]
  [AS select_statement];
```

Where row_format can be the following.

```
DELIMITED [FIELDS TERMINATED BY char [ESCAPED BY char]]
        [COLLECTION ITEMS TERMINATED BY char]
        [MAP KEYS TERMINATED BY char]
        [LINES TERMINATED BY char]
        [NULL DEFINED AS char]
```

And file_format is shown in the following.

```
SEQUENCEFILE  | TEXTFILE    -- (Default, depending on hive.default.
fileformat configuration)
              | RCFILE| ORC
              | PARQUET| AVRO
              | INPUTFORMAT input_format_classname OUTPUTFORMAT output_
                format_classname
```

Internal Tables

Internal tables are managed by Hive, and hence, called a *managed table*. To be managed by Hive means that when a user deletes a table, the data associated with it is also deleted from source files, along with the metadata. By default, Hive stores tables at /hive/warehouse. You can change this per your needs.

To understand how internal tables work, upload the movie data that you used at the start of this chapter to the /sample/movie_data/movie_metadata.csv location.

Now, create an internal table (without specifying an external keyword) using the following query. Here you are using the same movie data that you used previously.

```
CREATE TABLE movies
(
        director STRING, duration INT, gross DECIMAL,
        genres STRING, title STRING, language STRING,
        country STRING, ratings STRING, budget DECIMAL,
        release_year INT, score FLOAT
)
ROW FORMAT DELIMITED
FIELDS TERMINATED BY ','
LINES TERMINATED BY '\n'
STORED AS TEXTFILE;
```

Here you have not specified the location; hence, Hive creates a table in the default location at wasb://hive/warehouse. You can open Azure Storage Explorer and check the path. You'll find a new file named movies in there. It is a 0-bytes file. Next, you load data from the /example/movie_data folder. This is the Hive query for it:

```
LOAD DATA INPATH '/example/movie_data' INTO TABLE movies;
```

Executing the preceding query takes all the files from /example/movie_data folder and moves them into the /hive/warehouse/movies folder. Use any of the Azure Storage Explorer tools to verify this. Also, you can execute the DESCRIBE EXTENDED movies; query to show complete metadata about the table.

Now let's remove this table. Because it is an internal table, Hive also removes all the files associated with it. To drop a table, run the following query.

```
DROP TABLE movies;
```

At this point, if you open the Azure Storage Explorer tool and navigate to the /hive/warehouse folder, you won't see the movies folder anymore. That means you have completely lost the files inside movies folder, from the source and the Hive table folders. Hence, it is not advisable to create an internal table on a master copy of your data.

External Tables

External tables, opposite internal tables, do not copy your source files anywhere. Instead, Hive only stores table metadata. When a user submits a query on an external table, the data is read from the source location directly. That implies that when you drop am external table, Hive only removes the stored metadata and the data is kept intact for later use.

To see how an external table works, let's create one. Again, you are using the same data from the previous section. Upload the movies data to the /example/movie_data folder and create an external table on it, as shown in the following.

```
CREATE EXTERNAL TABLE movies
(
        director STRING, duration INT, gross DECIMAL,
        genres STRING, title STRING, language STRING,
        country STRING, ratings STRING, budget DECIMAL,
        release_year INT, score FLOAT
)
ROW FORMAT DELIMITED
FIELDS TERMINATED BY ','
LINES TERMINATED BY '\n'
STORED AS TEXTFILE
LOCATION '/example/movie_data/';
```

Note that you are specifying the location from which to load data into the table. Actually, Hive does not need to load data at all; it just reads it when the user submits a query. Now if you drop this table, your source files won't be affected. Run a drop query and verify that the files are inside the /example/movie_data/ folder; it won't be affected at all.

■ **Note** The TRUNCATE TABLE command is only applicable to internal tables, because Hive does not manage data in external tables and cannot delete it.

Storage Formats

Hive supports built-in and custom-developed file formats. The following are few of the built-in formats supported by Hive.

- **STORED AS TEXTFILE**: The default file format unless the hive. default.file format has been changed. As the name suggests, it stores as plain text files.

- **STORED AS SEQUENCFILE**: Sequence files are flat files, with key-value pairs in binary. It is compressed and it is the basic file format that Hadoop provides. Not only Hive but other tools in the Hadoop ecosystem also support the sequence file.

- **STORED AS ORC**: An optimized row columnar (ORC) format reduces the size of data up to 75% of the source file. It takes less space to store and less time to retrieve back, improving Hive query performance. It takes a small performance hit while updating data, but compensates this by providing a very high reading speed.

- **STORED AS PARQUET**: Another columnar storage format that Hive can work with. If your query only returns a few of the columns, then a columnar file format improves performance significantly.

- **STORED AS AVRO**: An Avro file format optimizes space. Hive natively works with Avro files. Avro is a binary format that stores a schema along with the data, allowing Hadoop to split a large file easily.

- **STORED AS RCFILE**: A binary storage format optimized specifically for a table with a large number of columns. This format splits data horizontally and then vertically. That means that it first divides data into row groups, and then stores each row group by dividing them vertically based on columns. This way, the row groups can be scanned in parallel, improving query performance. On top of that, if a query retrieves only a few columns, then it scans those columns only, further improving overall query performance.

Row Formats and SerDe

Serializer and deserializer (SerDe) in Hive allows you to read and write custom formats of data. Assume that you have an XML file that you want Hive to read and write. To tell Hive how to read/write this XML file, you need to specify a deserializer and a serializer, respectively, which converts XML to an object, and vice versa, for Hive to work with.

SerDe provides a way to convert the bytes stored in a file into a record and vice versa. In Hive, there are several built-in SerDes available: Avro, ORC, CSV, RegEx, and TSV. You can write your own custom SerDe and for formats not processable by the out-of-the-box SerDe available in Hive.

Partitioned Tables

When you have too much data in one table, query performance degrades as more and more data comes into the table. Also, if you only need part of the data, which can be a different block in a different query, then you can use the `PARTITIONED BY` clause to divide your table into multiple partitions that hold similar kinds of data in each block. To understand this concept, take the example of a log data file generated by a web server. In most cases, you only need the current month's data. If you have one large table with years of data in it, and you query only one of the months, Hive still has to scan all the data, which is not good for the user because this takes more time. What if Hive could automatically scan only a particular month of data and return the results? This automatic part can be implemented with the help of partitioned tables. In a partitioned table, you can divide your table based on one or more columns.

When you create a partitioned table, Hive internally creates folders for each partition key. If you use more than one column, then Hive generates a folder inside the folder; for example, if a table is partitioned by year and month columns, then you have a folder named partitionyear=2016, and inside that, you have folders partitionmonth=1 to 12. This way, when you query for December 2016, Hive knows that it only has to scan folder ../partitionyear=2016/partitionmonth=12, which improves the query performance significantly without adding any overhead to the client application. Clients can keep using the same query that they used before partitioning.

Let's try to create a simple partitioned table on the movies data based on release year.

```
CREATE TABLE movies_p
(
        director STRING, duration INT, gross DECIMAL,
        genres STRING, title STRING, language STRING,
        country STRING, ratings STRING, budget DECIMAL,
        score FLOAT
)
PARTITIONED BY (release_year INT)
ROW FORMAT DELIMITED
FIELDS TERMINATED BY ','
LINES TERMINATED BY '\n'
STORED AS TEXTFILE;
```

Note that you are not adding a release_year in the column list, but adding it to the PARTITIONED BY clause. To load data into a partitioned table, use the following query. Note that the following query assumes that you already have a movies table with data in it.

```
INSERT OVERWRITE TABLE movies_p PARTITION (release_year)
SELECT director, duration, gross, genres, title, language, country, ratings,
budget,score, release_year
FROM movies
WHERE release_year >= 2000 AND release_year <= 2016;
```

After executing the preceding statement, you should see a number of folders created in the /hive/warehouse/movies_p folder, as shown in Figure 4-11. All of these correspond to a year. And now you should query according to year only to optimize performance, like in the following query.

```
SELECT * FROM movies_p WHERE release_year = 2016;
```

Figure 4-11. *Hive partitioned table*

Furthermore, Hive can divide a table or partitions into buckets by using the CLUSTER BY clause on a column. Also, by using the SORT BY column clause, you can improve the performance significantly.

Create Table Options

Often with Hive, you want to create intermediate tables from the result of another query. In such cases, you can use the AS SELECT clause to create a table from the result of a query. Tables created using the AS SELECT clause are atomic, and cannot be seen by other users until the query results are populated. That is, other users either see the table with the complete results of a query or do not see the table at all.

Any select statement can be used with the AS SELECT clause, provided it is a valid HiveQL select statement. To copy data from a select statement, the target table should have matching table schemas. Please note that the AS SELECT clause cannot be used to create partitioned or external tables. The following is a sample snippet to create a table from another HiveQL query.

```
CREATE TABLE newtable
SELECT * FROM movies;
```

If you don't want to copy data but only the schema of a table to another table, then you can use the LIKE clause with the CREATE TABLE. The following is sample usage:

```
CREATE TABLE newtable LIKE movies;
```

Temporary Tables

Often, you need an intermediate table for the given calculation. In such scenarios, you can create temporary tables by using the TEMPRORARY clause with CREATE TABLE. These tables are only visible in the current session. They are deleted when the session ends. These tables don't support creating indexes or partitions on them.

In case there is already a table in the database with the same name as the temporary table, then for the current session, any reference to the table is resolved to the temporary table only. If a user wants to access a permanent table, then the temporary table should be dropped or renamed.

Data Retrieval

I executed select statements in previous sections but I did not provide a complete guide on what is available with the select statement. This section introduces different clauses and statements that can be used with the select statement to retrieve data.

The select statement is used to retrieve data from a Hive table. It can be used to filter, group, sort, and limit data coming out of Hive. The following is the syntax.

```
SELECT [ALL | DISTINCT] select_expr, select_expr, ...
  FROM table_reference
  [WHERE where_condition]
  [GROUP BY col_list]
  [HAVING expression]
  [ORDER BY col_list]
  [CLUSTER BY col_list| [DISTRIBUTE BY col_list] [SORT BY col_list]]
  [LIMIT number];
```

The following explains each clause and term. Also, everything that appears inside square brackets [ ]is optional.

- ALL or DISTINCT: Tells the query to return all rows or only distinct rows for the specified columns, respectively. If omitted, ALL is taken as the default value.

- table_reference is the input for the query. This can be table, view, join, or subquery.

- WHERE clause: Evaluated on each row. Query result contains all the rows where clause evaluates to true.

- GROUP BY clause: Combines same column values from different rows into single row and by applying aggregation function (e.g., sum, count) on different value columns to combine them.

- HAVING clause: Support for it was added in version 0.7.0. If you don't want to use the having clause, then you can use subquery to get the same results.

- ORDER BY clause: Sorts query results by one or more column values. It sort in ascending or descending order.

- LIMIT clause: Limits the number of rows returned by the query. No matter how many rows satisfy your query, it only returns a number specified in the limit clause. If the rows satisfying your query are less than the limit, then only those rows are returned.

- SORT BY clause: Similar to the order by clause, you can specify this clause, which instructs the engine to sort data by the column name before feeding it to a reducer.

- DISTRIBUTE BY clause: Distributes the rows among the reducers.

- CLUSTER BY clause: It is a combination of sort by and distribute by clauses.

- JOIN: Hive supports many different joins (i.e., inner join, left/right/full outer join, semi join, cross join, etc.), which covers most of the cases. A simple inner join can be written as shown in the following query:

  ```
  SELECT a.* FROM a JOIN b ON (a.id = b.id)
  ```

- UNION clause: Hive also supports the Union clause, which you can use to merge output from different select statement results. The following shows the syntax for UNION:

  ```
  select_statement UNION [ALL | DISTINCT] select_statement UNION
  [ALL | DISTINCT] select_statement
  ```

Note that this is not an exhaustive list of all clauses. HiveQL is expanding at a very rapid pace. There are many new clause supports coming. As of now, most of the traditional SQL stuff is already available.

Hive Metastore

In HDInsight, Hive internally uses the Azure SQL database to store schema information and other metadata. You won't get access to this SQL database unless you point it to your SQL database in Azure while creating a cluster. Normally, when you decommission your cluster, the associated SQL database is also deleted, and you lose your table structure. If you want to persist this structure, then you can point it to your SQL database. Note that your SQL database has to be in the same region as your HDInsight cluster. This way, you can be sure that whenever you spin up a cluster with an existing SQL database, you get your Hive structure back.

So, in a production environment, you might have hundreds of tables in Hive and you don't want to spin them up whenever you create a cluster for data processing. In such cases, you just point your cluster Hive metadata store to the existing Azure SQL database where you have already defined the structure for all of your Hive tables. And every time your HDInsight cluster comes up, it's pointing to that Azure SQL database to pull out the Hive metadata.

Apache Tez

Apache Tez, developed by Hortonworks, is built on top of YARN to provide high-performance data processing. Tez is a Hindi word that means *fast* —and Apache Tez truly justifies the meaning of it. It is an extensible framework for building high-performance batch and interactive data processing applications. Tez improves that performance of MapReduce while maintaining that ability to scale to petabytes of data.

As Hadoop was picked up by different companies, use cases started to expand. People wanted the near real-time output from Hive or Pig queries. But MapReduce cannot offer such high performance in near real-time, not because it is implemented incorrectly, but the way it works hinders MapReduce performance. Assume that you are making a Hive query that generates multiple MapReduce jobs and executes them one after another. Now after each reduce, intermediate data needs to be stored back to HDFS so that the next mapper can take it for processing. And due to multiple MapReduce jobs in succession, data needs to be written multiple times to disk. A disk write operation is quite slow, which causes bottlenecks to the whole query performance.

Tez solved this by removing the intermediate map operation and data streaming directly from one reducer to another. This significantly improved the whole query performance. A graphical representation of how MapReduce and Tez execute a query is shown in Figure 4-12.

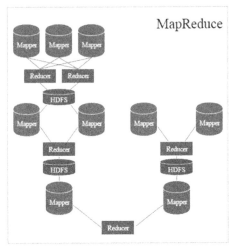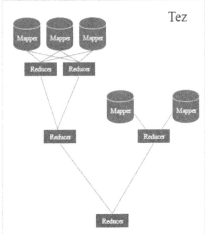

Figure 4-12. *MapReduce vs. Tez execution engine*

To get a better perspective of how fast Tez is, in the beginning of the chapter, I executed the SELECT COUNT(*) FROM movies; query using a MapReduce execution engine. And the result came in at around 42 seconds. In Figure 4-13, I executed the same query, but this time using the Tez execution engine. You can see that the performance of the query only took 3.7 seconds. Now you can imagine how much of a performance gain complex queries have.

```
hive> SELECT COUNT(*) FROM movies;
Query ID = sshuser_20161209125020_275705d3-2d00-4887-a91e-93e3463ab633
Total jobs = 1
Launching Job 1 out of 1


Status: Running (Executing on YARN cluster with App id application_1481280591580_0009)


--------------------------------------------------------------------------------

--------------------------------------------------------------------------------
Map 1 ..........   SUCCEEDED    1       1       0       0       0       0
Reducer 2 ......   SUCCEEDED    1       1       0       0       0       0
--------------------------------------------------------------------------------
VERTICES: 02/02  [==========================>>] 100%  ELAPSED TIME: 1.67 s
--------------------------------------------------------------------------------
Status: DAG finished successfully in 1.67 seconds


METHOD                     DURATION(ms)
parse                                 1
semanticAnalyze                     480
TezBuildDag                         263
TezSubmitToRunningDag                60
TotalPrepTime                     1,849

VERTICES        TOTAL_TASKS  FAILED_ATTEMPTS KILLED_TASKS DURATION_SECONDS   CPU_TIME
Map 1                    1                0            0             0.41
Reducer 2                1                0            0             0.44
OK
5044
Time taken: 3.736 seconds, Fetched: 1 row(s)
```

Figure 4-13. *Hive query with Tez execution engine*

Apache Tez creates a data flow graph to represent computation, where graph vertices represent application logic and edges represent data movement. Tez sits on top of YARN and provides execution for batch processing, Pig and Hive query execution, and other Hadoop components. In a distributed environment, estimating optimal data movement methods in advance is difficult. At runtime, as more information becomes available, it optimizes the execution plan further. YARN provides resources to the Tez engine based on available capacity and load at that point on the cluster. Also, it reuses every component in the pipeline so that operations are not duplicated unnecessarily.

There are two main benefits of Tez:

- **Cost-based optimization (CBO)**: The CBO engine uses statistics of tables and columns stored in Hive to generate an optimal query plan, which essentially increases the performance of the query and ensures better cluster utilization.

- **Vectorized query**: Vectorization enables data to be fetched in chunks, which means that it reads data in batch of 1000 rows at a time, compared to the usual 1 row at a time. This increases performance where a query scans through a large number of rows. This feature can be enabled from Ambari and only works with Hive tables in an optimized row columnar (ORC) file format.

Connecting to Hive Using ODBC and Power BI

So far, you have seen usage of command-line tools, web UI, and code-based interaction with Hive. But Hive can be used with Excel, Tableau, Power BI, and other similar tools through ODBC/JDBC connectors. In this section, you concentrate on ODBC connectors only.

ODBC and Power BI Configuration

To configure an ODBC connector for any of the tools, you need to download and configure it first. To download the Apache Hive ODBC driver for any client OS, go to `https://hortonworks.com/downloads/#data-platform` and locate the Hortonworks ODBC driver for Apache Hive in the Hortonworks Data Platform Add-Ons section, as shown in Figure 4-14.

Hortonworks Data Platform Add-Ons

Hortonworks ODBC Driver for Apache Hive (v2.1.5)

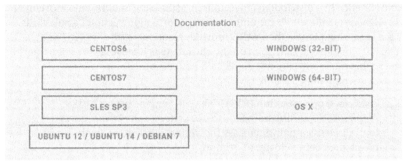

Figure 4-14. *Hortonworks ODBC driver for Apache Hive download section*

The following configures the Windows 64-bit version. Follow these steps to install the ODBC driver.

1. Depending on the bits needs of your client application, double-click to run HortonworksHiveODBC32.msi or HortonworksHiveODBC64.msi. And then click Next.

2. Select the checkbox to accept the terms of the license agreement, and then click Next.

3. To change the installation location, click Change, browse to the desired folder, and then click OK. The default should also work.

4. To accept the installation location, click Next. Then click Install.

5. When the installation completes, click Finish.

6. Verify that you have installed it correctly by going into the ODBC Data Source Administrator, System DSN tab. You should see the Hortonworks Hive DSN with the driver name Hortonworks Hive ODBC driver in it, as shown in Figure 4-15.

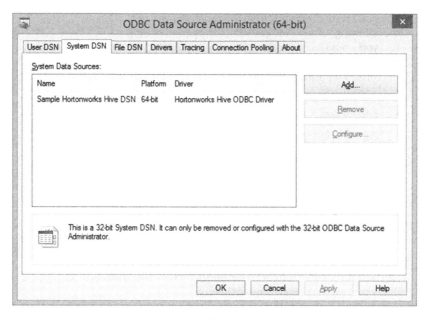

Figure 4-15. *Hortonworks Hive ODBC driver*

Now you need to configure the data source as per our cluster values if you plan to use something other than Power BI for data querying. Since you are using Power BI, you can skip configuration because you are providing a connection string instead. So, let's download and install Power BI.

1. To download, go to `https://powerbi.microsoft.com/en-us/get-started/` and download Power BI Desktop for Windows.

2. Once the installer is downloaded, open it and follow the installation instructions.

Prepare Data for Analysis

We will be analyzing flight delay data. The objective of the analysis is to identify airports with maximum delays (arrival and departure). The data that you are using comes from the Bureau of Transportation Statistics, (`www.transtats.bts.gov/DL_SelectFields.asp?Table_ID=236&DB_Short_Name=On-Time`). The US Bureau of Transportation Statistics collects data on the performance of major airline carriers that operate domestic flights, including departure delays and arrival delays. Follow this next procedure to get the required data.

1. Browse `http://www.transtats.bts.gov/DL_SelectFields.asp?Table_ID=236&DB_Short_Name=On-Time`.

97

2. You don't want everything from that page, so just select the following values. (You can select any period of time that you would like to work with. It is not mandatory to download the same three months. The same is available at the GitHub repository at `https://github.com/vinityad/airlinedelays`).

- **Filter Year**: 2016

- **Filter Period**: January, February, and March (download each month)

- **Fields**: Year, FlightDate, AirlineID, OriginAirportID, DestAirportID, DepDelay, and ArrDelay.

3. Save each month's data as January.csv, February.csv, and March.csv in one folder on your local machine. Collectively, all three files have around +1.3 million records.

4. You also need to download airlines and airports data, which can be downloaded from the repository at `https://github.com/vinityad/airlinedelaysGitHub`.

5. Upload all the folders—airlines, airports, and flight delay data—to Azure Blob storage in the `/2016_flight_delay` folder.

6. The final results in Azure Blob storage look like what's shown in Figure 4-16.

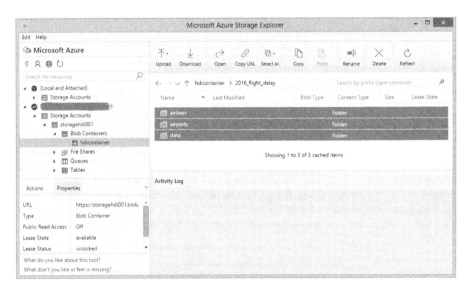

Figure 4-16. *Flight delay data in blob storage*

Creating Hive Tables

You are creating one external table for each dataset pointing to each of the three folders by using the following queries:

```
CREATE EXTERNAL TABLE flight_delays
(
    YEAR int,
    FL_DATE string,
    AIRLINE_ID int,
    ORIGIN_AIRPORT_ID string,
DEST_AIRPORT_ID string,
    DEP_DELAY int,
    ARR_DELAY int
)
ROW FORMAT DELIMITED FIELDS TERMINATED BY ','
LINES TERMINATED BY '\n'
STORED AS TEXTFILE
LOCATION '/2016_flight_delay/data'
TBLPROPERTIES ("skip.header.line.count"="1");

CREATE EXTERNAL TABLE airlines
(
    AIRLINE_ID int,
    Description string,
    Code string
)
ROW FORMAT DELIMITED FIELDS TERMINATED BY ','
LINES TERMINATED BY '\n'
STORED AS TEXTFILE
LOCATION '/2016_flight_delay/airlines'
TBLPROPERTIES ("skip.header.line.count"="1");

CREATE EXTFRNAl TABLE airports
(
    ID int,
    City string,
    Code string,
    Name string
)
ROW FORMAT DELIMITED FIELDS TERMINATED BY ','
LINES TERMINATED BY '\n'
STORED AS TEXTFILE
LOCATION '/2016_flight_delay/airports'
TBLPROPERTIES ("skip.header.line.count"="1");
```

Once you have all the tables, it is time to open Power BI and fetch the data in it.

Analyzing Data Using Power BI

To work with Hive, you need to download the Power BI desktop application. Power BI is a tool through which you can quickly analyze data from many different sources. The Power BI desktop tool can be downloaded from Power BI official website at `https://powerbi.microsoft.com`. Once downloaded, follow the standard install instructions.

Our final goal is to visualize the flight delay data and then figure out which airports to avoid based on delays. To do so, follow this procedure.

1. Open Power BI Desktop and click the Get Data button. This opens the Get Data window on which you can search for ODBC data source. Select it.

2. In the next window, select Sample Hortonworks Hive DSN as the data source name. Add the following connection string in the advanced options, as shown in Figure 4-17: `Driver={Hortonworks Hive ODBC Driver}; Host=hdi1.azurehdinsight.net; Port=443;Schema=default; RowsFetchedPerBlock=10000; HiveServerType=2; AuthMech=6; DefaultStringColumnLength=200;`.

3. Let's look at individual items in the connection string.

   - **Driver**: The type of driver. Since you are working with Hortonworks, the value is Hortonworks Hive ODBC Driver.

   - **Host**: Your Azure HDInsight cluster URL, or {ClusterName}.azurehdinsight.net.

   - **Port**: 443 for a secure connection.

   - **Schema**: The database that you want to connect with. In this case, you want to work with the default database only.

   - **RowsFetchedPerBlock**: The number of rows fetched as a block to improve performance. The default value is 10,000.

   - **HiveServerType**: You are working with Hive Server 2; hence, the value is 2.

   - **AuthMech**: The authentication mechanism to use; 6 means Windows authentication. 0: No Auth, 1: Kerberos, 2: Username, 3: Username and Password, 4: SSL Username and Password, 5: Windows Azure HDInsight Emulator, 6: Windows Azure HDInsight Service, 7: HTTP authentication, 8: HTTPS, 9: Kerberos over HTTP, and 10: Kerberos over HTTPS.

   - **DefaultStringColumnLength**: A string column the length of the Hive table's column. The default value is 32767, but you are using 200, because you know that you won't have more than 200 characters in a string column. Lower is better here.

Figure 4-17. *Power BI ODBC data source configuration*

4. After you click OK on the ODBC data source screen, you are presented with all the databases and their tables (when asked for credentials, enter your cluster credentials). You can select the tables that you want to include in the project. In our case, you select three tables—airlines, airports, and flight_delays, as shown in Figure 4-18. The navigator window not only shows you tables but also previews of the data stored inside selected tables. In Figure 4-18, you can see a few columns and rows from the flight_delay table.

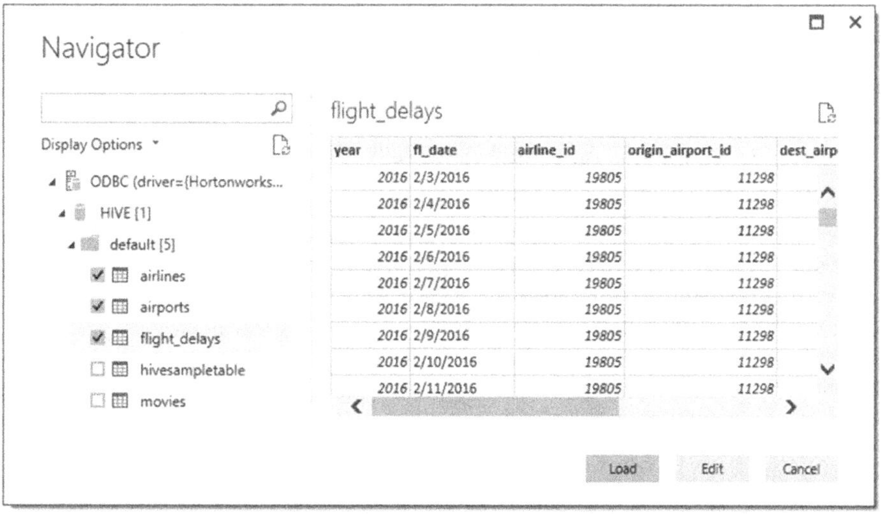

Figure 4-18. *Hive table selection in Power BI*

101

5. Clicking Load on the Navigator window instructs Power BI to load data from your HDInsight cluster. It may take some time, depending on the Internet connection speed you have, but it should not take long. Once it is ready with data, Power BI shows three tables in the fields section on right-hand side. It has also detected the types of fields and added a sigma symbol against numeric fields.

6. Now you need to create a relation between these tables. Use the Manage Relationship button from the ribbon to create a relation, as shown in Figure 4-19.

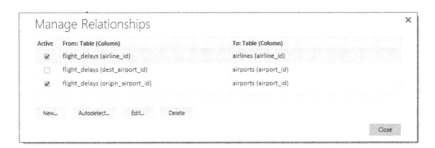

Figure 4-19. *Data relationship builder*

7. Start by creating a new column in the flight_delays table to identify delayed flights. Assuming that a flight departure delay of more than 30 minutes is a delayed flight, create a new column named is_flight_delayed.

8. To create a new column, click the three dots on the flight_ delays table ä New Column; or from the ribbon menu, Home ä New Measure ä New Column. Enter is_flight_delayed = flight_delays[dep_delay] > 30 in the column creation box that appears below the ribbon menu. This should create a new column of type Boolean in the flight_delays table.

9. To make a simple bar chart from the data, drag any bar chart from the Visualizations panel.

10. As soon as you select the bar chart, you see different placeholders for the chart in the Visualizations panel below all the charts, as shown in Figure 4-20.

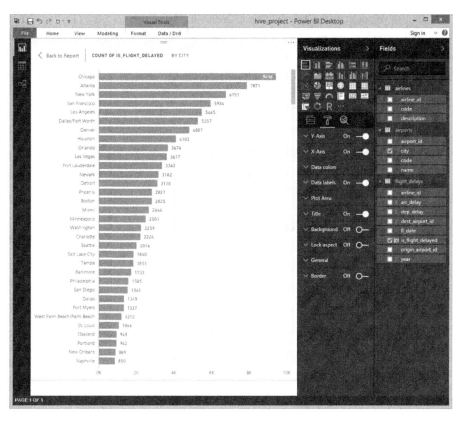

Figure 4-20. *Flight delay bar chart visualization*

11.   First drop the city from the airports table on the Axis placeholder. Then drop is_flight_delayed to the Value placeholder. Note that Power BI is smart enough to convert the value field to Count of is_flight_delayed.

12.   Now if you look at the data, it is not correct. Atlanta shows around 90K records, which can't be true. If you look at the value field, it is counting everything, whether true or false; hence, you should filter whole page data to only include delayed flights. To add a page level filter, drag the is_flight_delayed field in the Filters section under the Page level filters placeholder. And then select only the True value in it, which gives data for only delayed flights. Chicago comes in first with the most delayed flights, at around 9.4K.

13. If you change the Axis column to the Airline's description field, which is actually airline name (you can rename it by right-clicking the column name). This changes the bar chart, as shown in Figure 4-21. You can immediately see which airline to avoid if you don't want to be delayed.

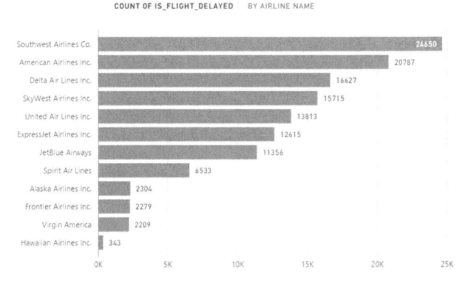

COUNT OF IS_FLIGHT_DELAYED   BY AIRLINE NAME

Figure 4-21. *Most number of delays by airlines*

14. But as I said before, you want to visualize the data on map and not some boring bars. Open a new page, using the plus icon in the bottom-left corner.

15. On the new page, put a new Map visualization (globe icon button).

16. In this visualization, you need to provide the city from airports to Location placeholder field.

17. Add the is_flight_delayed column to the Size placeholder.

18. Now, add is_flight_delayed to the Page level filter and select only the true value.

19. You are interested in only the top few cities with the most departure delays; hence, add a filter to City, select the filter type as Top N, and type **20** in the value field.

20. The final result of the procedure should look like what's shown in Figure 4-22.

COUNT OF IS_FLIGHT_DELAYED   BY CITY

Figure 4-22. *Most delayed departures by city*

Hive UDFs in C#

Hive is a great tool to work with HDInsight. But what if you want to do something that is not possible to write as HiveQL or is not yet supported by it. Also, many times you need more general-purpose methods in your query. Apache Hive solves both of these issues by providing User Defined Functions, which can be written in a variety of languages, including Java and C# (HDInsight cluster).

There are three types of functions in Hive.

- **UDF**: A regular User Defined Function. When you cannot write logic using built-in Hive functions.

- **UDAF**: User Defined Aggregate Function. Works on more than one row to produce aggregated output, like the Count or the Sum built-in functions.

- **UDTF**: User Defined Tabular Function. Works in the reverse way. It takes one row as input and returns multiple rows as output. For example, in the movies table, you have a genre column, which is pipe (|) delimited. You can use EXPLODE to get that data as rows, using a query such as this:

```
SELECT EXPLODE(genres) FROM (SELECT split(genres, '\\|') as genres FROM
movies LIMIT 1) genre_table;
```

It returns the row Action|Adventure|Fantasy|Sci-Fi as four rows, each having a single genre value.

User Defined Function (UDF)

You can create functions in C# and execute it in a Hive query, just like EXPLODE in the last query. Let's first create a simple program that takes an input and generates MD5 hash for it.

1. Create a new console application in C# and name it HiveUDF.

2. In your Main method, add the following code, which takes a line as input and returns an MD5 hash of the same.

```
class Program
{
    static void Main(string[] args)
    {
        string line;
        // Read stdin in a loop
        while ((line = Console.ReadLine()) != null)
        {
            // Parse the string, trimming line feeds and splitting fields at
            tabs
            var field = line.TrimEnd('\n');

            // Emit new data to stdout, delimited by tabs
Console.WriteLine("{0}\t{1}", field, GetMD5Hash(field));
        }
    }

    /// <summary>
    /// Returns an MD5 hash for the given string
    /// </summary>
    /// <param name="input">string value</param>
    /// <returns>an MD5 hash</returns>
    static string GetMD5Hash(string input)
    {
        // Step 1, calculate MD5 hash from input
        MD5 md5 = System.Security.Cryptography.MD5.Create();
        byte[] inputBytes = System.Text.Encoding.ASCII.GetBytes(input);
        byte[] hash = md5.ComputeHash(inputBytes);

        // Step 2, convert byte array to hex string
        StringBuilder sb = new StringBuilder();
        for (int i = 0; i < hash.Length; i++)
        {
            sb.Append(hash[i].ToString("x2"));
        }
        return sb.ToString();
    }
}
```

3. Compile it and upload the .exe file from the debug folder to Azure Blog storage for the HDInsight cluster in a folder called UDFs.

4. Use this .exe (UDF) in a Hive query. Fetch the title from the movies table and generate a hash of them. A query with UDF looks like the following:

```
SELECT TRANSFORM (title) USING 'wasbs:///UDFs/HiveUDF.exe' AS (title string, hash string) FROM movies LIMIT 5;
```

Here you are taking a title from the movies table. You know that our method (UDF) returns tab-separated fields by title because it is the first field and the MD5 hash is the second field. This output is then returned on the console. Figure 4-23 shows the output of the preceding query (the execution part has been removed for brevity).

Figure 4-23. *User-defined function execution*

User Defined Aggregate Functions (UDAF)

User Defined Aggregate Functions take multiple rows and return a single/aggregate row as output. Let's look at how this is done in a C# program.

1. Create a new console application in C# and name it HiveUDAF.

2. In your Main method, add the following code, which takes all the genres as input and generates an aggregate output of the genres count.

```
static void Main(string[] args)
{
    string line;
    List<string> genres = new List<string>();
```

```
    // Read stdin in a loop
    while ((line = Console.ReadLine()) != null)
    {
        // Parse the string, trimming line feeds and splitting fields at tabs
        var field = line.TrimEnd('\n');

        // Saperate the genres into list
        genres.AddRange(field.Split('|'));
    }

    var distinctGenres = genres.Distinct().ToList();

    foreach (var genre in distinctGenres)
    {
        Console.WriteLine($"{genre}\t{genres.Count(o => o.Equals(genre))}");
    }
}
```

3. Compile it and upload the .exe file from the debug folder to Azure Blog storage of the HDInsight cluster in a folder called UDFs.

4. Use this HiveUDAF.exe (UDF) in a Hive query. Fetch genres from the movies table and count the total number of times they appear in the list. The query with UDF looks like the following:

```
SELECT TRANSFORM (genres) USING 'wasbs:///UDFs/HiveUDAF.exe' AS (genres
string, count int) FROM movies ORDER BY count DESC LIMIT 5;
```

Here you are taking genres from the movies table. You know that our method (UDF) returns tab-separated fields for genres because it is the first field and counts it as the second field. This output is then returned on the console. Figure 4-24 shows the output of the preceding query (the execution part has been removed for brevity).

Figure 4-24. *User-defined aggregate function execution*

User Defined Tabular Functions (UDTF)

User Defined Tabular Functions convert a single row to multiple rows. Like the EXPLODE function in Hive, which takes an array and returns it as rows, you try to build a similar C# function and use it in Hive query.

1. Create a new console application in C# and name it HiveUDAF.

2. In the Main method, add following code, which takes all the genres as input and generates aggregate output of each genre count.

```
class Program
{
    static void Main(string[] args)
    {
        string line;

        // Read stdin in a loop
        while ((line = Console.ReadLine()) != null)
        {
            // Parse the string, trimming line feeds and splitting fields at
            tabs
            var field = line.TrimEnd('\n');

            // Saperate the genres into list
            foreach (var item in field.Split('|'))
            {
                Console.WriteLine(item);
            }
        }
    }
}
```

109

3. Compile it and upload the .exe file from the debug folder to the Azure Blog storage of the HDInsight cluster into a folder called UDFs.

4. Use this HiveUDTF.exe (UDF) in a Hive query. Fetch genres from the movies table and explode it into rows. A query with UDF looks like the following:

```
SELECT TRANSFORM (genres) USING 'wasbs:///UDFs/HiveUDTF.exe' AS (genres
string) FROM (SELECT genres FROM movies LIMIT 1) genre_table;
```

Here you are taking genres from the movies table. You know that our method (UDF) returns a list of genres. Figure 4-25 shows the output of the preceding query (the execution part has been removed for brevity).

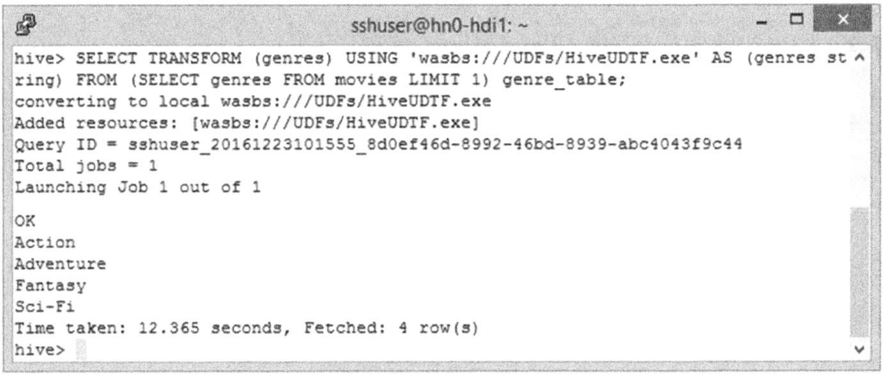

Figure 4-25. *User-defined tabular function execution*

Summary

In this chapter, you explored Apache Hive fundamentals. Hive makes it easy for developers, BI professionals, and SQL users to interact with Hadoop data. They can use their SQL knowledge to query the data without writing complex MapReduce jobs. Then you went through different ways to connect to Hive on an HDInsight cluster, which not only includes graphical and command-line tools, but also code-based approaches. As a graphical tool, Power BI is an attractive option to quickly analyze and pattern discovery. Finally, you explored other advanced ways to query using User Defined Functions in Hive.

Overall, Hive is a complete package for working with Big Data projects. You can build almost everything related to querying with Hive.

The next chapter takes a look at another SQL-like tool: Apache Pig.

CHAPTER 5

▩ ▩ ▩

Using Pig with HDInsight

Apache Pig is a platform to analyze large data sets using a procedural language known as Pig Latin. One of the challenges with MapReduce is that to represent complex processing, you have to create multiple MapReduce operations and then chain them together to achieve the desired result, which is not easy or maintainable when requirements change very often. Instead, you can use Pig, which represents transformations as a data flow. You can write different transformations, one after another, to achieve the desired result. Apache Pig is mainly used in data manipulation operations, because it is easier to write in Pig Latin than to write basic MapReduce jobs in Java. Pig Latin is the language used by Pig to write procedures to do transformations. Pig Latin procedures usually consist of one or more operations, such as loading data from a file system, manipulating it, and storing the output for processing or dumping it on a screen.

Apache Pig's main advantage is the capability to substantially parallelize, which allows it to handle very large data sets. Other advantages of Apache Pig and Pig Latin are as follows.

- **Easy to program**: Programmers not comfortable with writing low-level MapReduce in Java (or similar) languages can work with Hadoop using comparatively easy to write SQL, such as the Pig Latin language. Complex tasks can be easily encoded as data flow sequences. Pig procedures can do large amounts of work, yet they are easy to write and maintain.

- **Extensible**: Pig allows users to write complex operations as a custom function called *user defined functions*.

- **Optimization**: Pig automatically optimizes submitted jobs, allowing users to focus on business tasks.

Pig is designed to work with long-running data operations, which makes it suitable for the following situations.

- Data extract-transform-load (ETL) jobs

- Analysis/research on raw data

- Iterative data processing

© Vinit Yadav 2017

V. Yadav, *Processing Big Data with Azure HDInsight*, DOI 10.1007/978-1-4842-2869-2_5

When you perform a transformation on data, you get a relation. In the next transformation, you use the resultant relation from the previous transformation. This way, you can chain together transformations such as filter, grouping, sorting, and so forth, which generates a relation as a result. Initially, relations are created using the Load command. It is also a schema on read, such as Hive. In Hive, when you create a table while in Pig, it is called a *relation*. Once the schema is created, it is overlaid on top of existing data. Like Hive, if there is a data-type mismatch, then Pig also puts null as the value. To run Pig Latin statements, there is a command-line shell called Grunt.

The Apache Pig architecture consists of a Parser, Optimizer, Compiler, and the execution engine. A Pig script goes through all of these components to generate output. A Pig script first goes through Parser, which checks for any syntactical errors, and performs type checking and other checks. It converts the Pig script to a directed acyclic graph, a logical plan in which nodes are the logical operator and edges are represented by data flow. Pig splits, merges, transforms, and reorders operations to optimize the plan. Then this logical plan is converted to a MapReduce plan. Here, the MapReduce job boundaries are defined. This plan is then compiled into a series of MapReduce jobs. And finally, MapReduce jobs are submitted to Hadoop in ordered fashion by the execution engine and output is generated in a file or in shell based on commands in the Pig script.

Understanding Relations, Bags, Tuples, and Fields

A Pig relation is similar to a table in relational databases. It is called an *outer bag*. A bag consists of one or more tuples, which you can think of as rows in a traditional database. It has multiple tuples, which represent the rows of data. Within each tuple, there are ordered sets of fields, which are actual values. But unlike a traditional database, Pig doesn't require you to have the same number of fields in every tuple. Relations are also unordered, which means that tuples may be processed without any particular order. Figure 5-1 shows a sample relation, tuples, and fields. Relation is represented by curly braces. A sample relation contains three tuples represented by round braces, and each tuple contains two fields.

```
{(a, 1)      • A relation is an outer bag
(b, 2)           ○ A bag is a collection of tuples
(c, 3)}          ○ A tuple is an ordered set of fields
                 ○ A field is a data item
```

Figure 5-1. *Sample relation, tuples and fields*

So far, a relation is totally identical to a relational database table. But things get interesting when you add another tuple with an inner bag in it. In Figure 5-2, the fourth tuple contains two fields: the first field's value is d, while the second one is an inner bag, which again contains two tuples and two fields each. Not only can tuples have inner bags, but also there can be a different number of fields in a tuple. You can have a non-matching schema of tuples in a relation.

In Figure 5-2, the fifth and sixth tuples have completely different numbers of fields compared to previous tuples. The fifth tuple has only one field while the sixth tuple has three fields. This is a very flexible structure to hold data. You can project a schema as per your data and not necessarily follow the same schema for every tuple, which makes it very powerful in ETL scenarios.

```
{(a, 1)
(b, 2)
(c, 3)
(d, {(4, 5), (6, 7)})
(e)
(f, 8, 9)}
```

- A field can contain an inner bag
- A bag can contain tuples with non-matching schema

Figure 5-2. *Tuples with different schema/fields*

Relations can be referenced using a name (or alias). To create a relation with a reference, use the following syntax. Please note that the name of the relation, field, function, and so forth, is case sensitive.

```
EMP = LOAD 'employee' USING PigStorage() AS (name:chararray,
address:chararray, phone:chararray, age:int, salary:float);
```

In the preceding code snippet, you are loading data into an EMP relation with five fields. Fields can be referenced by the name assigned by the schema or by using positional notation. Positional notation is system generated. It starts with 0 and is prefixed with a dollar $ sign (i.e., $0 refers to first field). In the preceding code snippet, it refers to the name field. Similarly, $1 refers to the address field, and so on. The following statement selects the name and age ($3) from the EMP relation.

```
EMP2 = FOREACH EMP GENERATE name, $3;
DUMP EMP2;
(Scott, 25)
(Bill, 40)
(Joe, 32)
```

Fields can have complex data types as well. In the next code snippet, there is the contact information schema. Each person can have multiple types of contact information, such as email, phone, fax, and so forth. There may be multiples of each type of contact information as well.

```
PERSON = LOAD 'data' AS (name: chararray, (contact: bag{c: tuple (type:
chararray, value: chararray)});
DUMP PERSON;
(Joe, {(email, joe@mail.com), (phone, 1231231234)})
(Scott, {(phone, 0987654321), (phone, 1122334455)})
```

Data Types

Apache Pig has both simple and complex data types. Tables 5-1 and 5-2 list all available data types. The simple and the complex data types are listed, respectively.

Table 5-1. *Simple Data Types*

Data Type	Description	Example
Int	Signed 32-bit integer	10
Long	Signed 64-bit integer	10L or 10l
Float	32-bit floating point	10.2F or 10.2f
Double	64-bit floating point	10.2 or 10.2e4 or 10.2E4
Chararray	Character array (string) in Unicode UTF-8 format	Hello World
Bytearray	Byte array (blob)	

Table 5-2. *Complex Data Types*

Data Type	Description	Example
Tuple	An ordered set of fields	(10, 1)
Bag	A collection of tuples	{(10, 2), (10, 3)}
Map	A set of key value pairs	[color#red]

Here are a few general observations about data types.

- To assign field types, use a schema. If a type is not specified, then the default type bytearray is used, and depending on the context, implicit conversions are applied. For example, in the following code, in relation Y, field a1 is converted to an integer, while in relation Z, a1 and a2 are converted to double, because you don't know the type of either fields.

    ```
    X = LOAD 'data' AS (a1, a2);
    Y = FOREACH X GENERATE a1 + 10;
    Z = FOREACH X GENERATE a1 + a2;
    ```

- If the load statement has a schema defined with it, then the load function tries to enforce it on the data; if any value doesn't conform to the schema, then a null value or an error is generated.

- If an explicit cast is specified and data cannot be converted, then an error is generated. For example, if the data is a chararray and the field type is an integer with an explicit cast, then it generates an error. In the following example, the second line generates an error.

```
X = LOAD 'data' AS (a1:chararray);
Y = FOREACH X GENERATE (int)a1;
```

Connecting to Pig

To execute a Pig script, you can connect to Pig using your choice of CLI utility. You can connect with PowerShell or SSH to a Linux cluster and then type **Pig** to get into the Grunt shell. Grunt shell is the shell used to work with Pig. Also, Pig supports .NET SDK in an HDInsight cluster. This allows you to integrate Pig scripts in your project directly without relying on external tools.

To connect, use SSH from the Linux host or use PuTTY from the Windows client machine (refer to the "Connecting to cluster using SSH" section in Chapter 2). Once connected, type **Pig**. You should see the grunt> prompt, which means that you are connected, and you can start writing Pig Latin statements and scripts. Figure 5-3 shows a screenshot of PuTTY connected to a Linux cluster and the Grunt shell started in it.

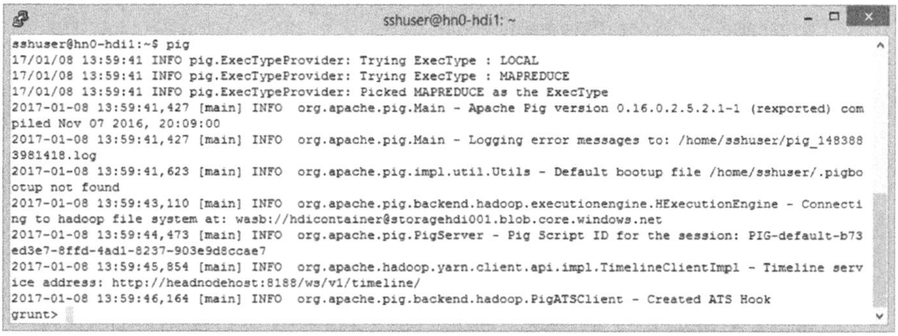

Figure 5-3. *Grunt shell on Windows host using PuTTY*

To integrate Pig into your project workflow through code, you can use .NET SDK. You saw how Hive integrates in .NET projects in Chapter 4. Submitting a Pig script should be very familiar. The following describes the method for submitting a Pig job.

1. Create a C# console application in Visual Studio (use Visual Studio 2012 or higher) targeting .NET Framework 4.5 or higher.

2. To get SDK bits, install a Nuget package for HDInsight using the Install-Package Microsoft.Azure.Management.HDInsight. Job command. This installs its dependencies as well.

115

HDInsightJobManagementClient is the main class that facilitates communicating with the HDInsight service. It comes from the Microsoft.Azure.Management.HDInsight.Job namespace. It requires an object of cluster credentials and a cluster Uri to instantiate it.

Important methods from the perspective of submitting a Pig job are SubmitPigJob, WaitForJobCompletion and GetJob in HDInsightJobManagementClient.

The following is the complete routine to submit a Pig script to fetch sorted airlines data by their description and then store it in the ordered_airlines folder.

```
private static void SubmitPigJob()
{
    // Cluster credentials
    Console.WriteLine("Enter cluster http credentils");
    Console.Write("Username: ");
    var clusterUsername = Console.ReadLine();
    Console.Write("Password: ");
    var clusterPassword = GetMaskedPassword();

    // Cluster name
    Console.Write("Enter cluster name: ");
    var clusterName = Console.ReadLine();
    var clusterUri = $"{clusterName}.azurehdinsight.net";

    var clusterCredentials = new BasicAuthenticationCloudCredentials()
    {
        Username = clusterUsername,
        Password = clusterPassword
    };
    HDInsightJobManagementClient jobManagementClient =
                new HDInsightJobManagementClient(clusterUri,
                clusterCredentials);

    //List<string> args = new List<string> { { "argA" }, { "argB" } };
    var parameters = new PigJobSubmissionParameters
    {
        Query = "airlines = LOAD '/2016_flight_delay/airlines/airlines.csv'
        USING PigStorage(',') AS (airline_id:int, description:chararray,
        code:chararray);" +
        "ordered = ORDER airlines BY description;" +
        "STORE ordered INTO '/2016_flight_delay/airlines/ordered_airlines'
        USING PigStorage(',');"
    };
    System.Console.WriteLine("Submitting the Hive job to the cluster...");
    var jobResponse = jobManagementClient.JobManagement.
    SubmitPigJob(parameters);
    var jobId = jobResponse.JobSubmissionJsonResponse.Id;
    System.Console.WriteLine("Response status code is " + jobResponse.
    StatusCode);
    System.Console.WriteLine("JobId is " + jobId);
```

```
System.Console.WriteLine("Waiting for the job completion ...");

// Wait for job completion
JobGetResponse waitResponse = jobManagementClient.JobManagement.
WaitForJobCompletion(jobId);

System.Console.WriteLine("Pig Job Completed.");
}
```

Please note that you are using the GetMaskedPassword method from the previous chapter. The job output is stored in Azure Blob storage in the /2016_flight_delay/ airlines/ordered_airlines folder.

Operators and Commands

So now, you know how to connect with Pig and start a Grunt shell. Next, let's explore that different operators and commands available to work with data. To do this, let's generate a list of the airlines that have the highest average number of delays in the month of January. You already have this data in Azure Blob storage from the previous chapter. Use airlines/airlines.csv and data/January.csv (if you don't have this data, then have a look at the "Prepare Data for Analysis" section in Chapter 4). The following are the transformations to carry out on the data.

1. Find all the records with departure delays higher than or equal to 30 minutes.

2. Find the average departure delay for each airline.

3. Sort the results by average delay in descending order, or from highest to lowest.

4. Show the top 10 airlines with the highest average departure delay.

Let's start by loading the data into a relation.

1. Load January's data into a relation called data, which contains the year, flight_date, airline_id, origin_airport, dest_ airport, departure_delay, arrival_delay tuples separated by commas. Here you are using PigStorage to specify it is a comma-delimited file. All the fields are of type integer except flight_date, which is a chararray. The airline_id field has the id that matches with records in the airlines CSV file.

```
data = LOAD '/2016_flight_delay/data/january.csv' USING
PigStorage(',') AS (year:int, flight_date:chararray,
airline_id:int, origin_airport:int, dest_airport:int,
departure_delay:int, arrival_delay:int);
```

117

2. Next, filter out records with departure delays less than 30 minutes. This prevents negative values and small delays from affecting the overall calculation.

```
filteredData = FILTER data BY departure_delay >= 30;
```

3. You have only the larger departure delay data now. The schema is still the same as when you loaded data, which contains multiple records for each airline, with different flights and their respective delays. Group this data by the airline_id field, which generates two tuples: airline_id and one containing a bag of tuples, the same as the original schema. Use filteredData relation from the previous step and generate a new relation, groupedData.

```
groupedData = GROUP filteredData BY airline_id;
```

4. Find the average departure delay of each grouped airline from the groupedData relation. To find the average, go through each record in the bag and then generate the average departure delay using the AVG inbuilt operator. After applying this step, you get a flattened schema with a bag of two tuples, airline_id, and the average departure delay of the airline. You are not interested in decimal points in the average delay; hence, explicitly cast the average departure delay to an integer. The Pig script is as follows.

```
groupedAvgs = FOREACH groupedData GENERATE group as
airline_id, (int)AVG(filteredData.departure_delay) as
avgDelay;
```

5. At this point, you have airlines' average delays in January; but you don't have the names of the airlines, just their ids. To get the name of an airline, you have to join the groupedAvgs result with the airlines data that you have in another CSV file. But before that, you need to load the airlines data into a relation. Let's call this relation airlines, and the tuples are airline_id, description(name), and the code for each airline.

```
airlines = LOAD '/2016_flight_delay/airlines/airlines.
csv' USING PigStorage(',') AS (airline_id:int,
description:chararray, code:chararray);
```

6. Join the `airlines` relation with the `groupedAvgs` relation. Using the `join` operator, you apply the join based on airline_id. This join gives you a new relation with all the tuples from both relations. This means that now you have airline_id (twice), description, code, and average departure delay in the `joinedData` relation.

    ```
    joinedData = JOIN groupedAvgs BY airline_id, airlines
    BY airline_id;
    ```

7. Because the processing was distributed across multiple jobs on multiple nodes, you don't have any guarantees about the order of the data; it's been distributed to different nodes and then reassembled into the relation. So, the next thing to do is sort the data by the average departure delay in descending order to have the airlines with the most delays at the top.

    ```
    sortedData = ORDER joinedData BY avgDelay DESC;
    ```

8. You have all the airlines in the sorted list with the highest average departure delay at the top. The final step is to fetch only the first ten rows from the data, which you can do using `limit` operator.

    ```
    top10AirlineDelays = LIMIT sortedData 10;
    ```

9. To see the data, use the `dump` operator to dump all the rows onto the screen in the `top10AirlineDelays` relation.

    ```
    DUMP top10AirlineDelays;
    ```

The complete script for the whole process is as follows.

```
// Load flight delay data
data = LOAD '/2016_flight_delay/data/january.csv'
        USING PigStorage(',') AS
        (year:int, flight_date:chararray, airline_id:int, origin_
airport:int,
        dest_airport:int, departure_delay:int, arrival_delay:int);

// Filter delays less than 30 minutes
filteredData = FILTER data BY departure_delay >= 30;

// Group data by airline
groupedData = GROUP filteredData BY airline_id;
```

```
// Flattern the grouped airlines and find average delay
groupedAvgs = FOREACH groupedData GENERATE group as airline_id,
        (int)AVG(filteredData.departure_delay) as avgDelay;

// Load airline names
airlines = LOAD '/2016_flight_delay/airlines/airlines.csv'
        USING PigStorage(',') AS (airline_id:int, description:chararray,
code:chararray);

// Join airlines names with flattern airline delays
joinedData = JOIN groupedAvgs BY airline_id, airlines BY airline_id;

// Sort data by highest delay to lowest
sortedData = ORDER joinedData BY avgDelay DESC;

// Fetch only first 10 record and show them on console
top10AirlineDelays = LIMIT sortedData 10;
DUMP top10AirlineDelays;
```

The result of the script is shown in Figure 5-4.

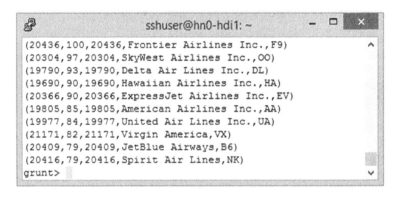

Figure 5-4. *Top 10 average departure delayed airlines for January 2016*

When executing each statement, you might notice that it returns to shell immediately. That is because Pig won't run any of the statement until you ask for the result using dump or store. So, as soon as you enter last dump statement in the script, it starts the MapReduce job based on the statements that you entered, and then shows the result.

You used quite a few operators and commands in the script, but there are a few more. The following is a list of the most commonly used operators and commands.

- The LOAD command instructs Pig to get data. It also tells the format in which it should be read and the location of the data. Often this is the entry point in the script.

- The FILTER command removes the rows that you are not interested in. In our sample, you removed the departure delays that were less than 30 minutes. In another scenario, you might want to remove the first row because it contains header text or remove blank fields. This leaves you with the rows that you are interested in working further with.

- FOR EACH ... GENERATE is a common operation in Pig. It allows you to iterate over the relation supplied after FOR EACH and performs operations on each row of data, generating new tuples from it.

- ORDER BY allows you to sort relations in a particular order, such as ascending or descending. Also, it supports multiple fields.

- JOIN allows you to combine two relations based on a common field. You can decide whether you want to keep the duplicated field value from both tables or if you just want to merge them into one.

- The GROUP command allows you to turn a flat tuple into a tuple and bag combination, effectively grouping similar tuple values into one tuple. You end up with a group-level tuple that contains a bag inside it. This bag contains all the tuples matching the grouped value.

- The FLATTEN command flattens out the nested tuple and inner bag, so you end up with just the fields.

- LIMIT is a way of restricting the number of tuples that you're getting in the result. So, if you have a large set of results but you only want the first ten tuples, then you can limit the results to a certain number, as you did with our script.

- DUMP removes the contents of a relation and displays it on the console, or dumps out straight to the output from where it is called. Normally, it is used when you are troubleshooting a Pig Latin script as a way to peek into the data generated after any step.

- STORE saves the result of a script into a file on a shared file system. It takes a relation and formats it to store (i.e., comma delimited), and then creates file(s) for it, which you can use further with other tools such as Hive for analysis.

Internally, all of these commands are translated into Map and Reduce jobs, which are submitted to Hadoop. In addition to the commands listed, there are a few more available in Pig Latin, including IMPORT, DISTINCT, RANK, SPLIT, STREAM, and UNION.

Executing Pig Scripts

You saw how to run Pig Latin statements line by line using the Grunt shell. It is a very common scenario initially, when you are exploring data. But typically, you bundle Pig Latin statements into a script and run it as one operation. So, let's take a look at how you run Pig scripts.

First, you need to have Pig script, so create a text file with the script from the previous section, which generates the top 10 airlines with maximum delays. Save it as airline. pig. You also need to upload it to the Azure Blob storage associated with the HDInsight cluster. Once you've saved the Pig Latin script, you can then run it using Pig, such as `pig wasb://scripts/airlines.pig` from the command line. Alternatively, you could use PowerShell or the Hue environment on the dashboard for your cluster. There are various ways to initiate a Pig job, but what you are effectively doing is running Pig and pointing it at the script file to run the job. Once you have run the job, it generates output in the form of a file or multiple files. Then you consume the results by using any of the clients, such as Excel, PowerBI, or anything else. Because you are using HDInsight, the results are stored in the shared storage in the Azure Blob store used by the cluster.

Summary

This chapter covered Apache Pig's fundamentals. Pig helps you build data transformation pipelines before the data is ingested into analytical tools. Use it to clean data, remove outliers, normalize, and group and sort data. In HDInsight, Pig directly takes input from Azure Blob storage and stores the results as a file(s) back into Azure Blob storage. The results can be given as input to other tools. As a .NET developer, you also get .NET SDK, through which you can easily integrate it in your project workflow. Overall, Apache Pig is an easy to use tool when it comes to data processing.

In the next chapter, you look at Apache HBase to store large amounts of data with high read/write throughput over Hadoop.

CHAPTER 6

■ ■ ■

Working with HBase

Previous chapters explored how to leverage an HDInsight cluster to store and process big data. You learned how MapReduce jobs process data. Also, you looked at Hive and Pig, and learned how they make it easy to work with data. All the technologies and tools that you saw so far work in batch mode. And they are accepted in *online analytical processing* (OLAP) scenarios where it is supposed to take time. But you cannot always use batch processing. What if you want a low-latency database that provides near real-time read/ write access, and quick random access to your big data in Hadoop? This is where Apache HBase comes into the picture.

Apache HBase is an open source, non-relational, distributed database. It is based on Google's Bigtable implementation and is written in Java. It is a NoSQL database that provides real-time read/write access to large data sets. In this chapter, you explore HBase: its architecture, processing data using HBase Shell, and leveraging .NET SDK to read/ write/query data on an HBase cluster inside HDInsight.

Overview

Apache HBase is a NoSQL database that runs on top of HDFS. With HDInsight, it utilizes Azure Blob storage as the default file system. HBase provides a querying interface over Hadoop, enabling access to data in near real time. HBase can handle millions of columns and billions of rows. It gets this ability from its scaling capability. It exhibits linear scalability. It has a very good automatic sharding model in which HBase automatically distributes tables when they become too large. In HBase, you can have multiple machines acting as a single unified system. And if you need more capacity in terms of storage or transactions throughput, you can add more machines and it scales accordingly.

NoSQL databases are not a new concept, but they weren't very popular at first. The data explosion made developers think about an alternative approach to traditional relational databases for storing and retrieving data. NoSQL databases use data structures like key-value pairs, wide columns, graphs, or documents. There are many different NoSQL databases; popular ones are MongoDB, Cassandra, Redis, and HBase. The suitability of a NoSQL database depends on the situation, as each database is designed to solve different problems.

© Vinit Yadav 2017

V. Yadav, *Processing Big Data with Azure HDInsight*, DOI 10.1007/978-1-4842-2869-2_6

HBase wias built to host very large tables with varying structures, makng it a good choice to store multi-structure or sparse data. HBase consists of tables, and a table contains columns and rows, much like a traditional relational database (RDBMS) table. HBase provides fast random access to data, and it achieves that by having a fast tree lookup. HBase is known for high write throughput; hence, HBase works very well for systems that require a lot of writes. Data in HBase is timestamped and multiple versions are maintained. This makes point-in-time and flashback queries possible. You can control the number of versions of your data that you want retained. Querying without specifying the version returns the most recent data only.

HBase is a columnar store database. That means data from a single column is stored together. This is completely opposite from relational databases, where rows are stored together. Columnar storage provides flexibility in the number of columns a row can have, making a table sparse. This makes it easy and inexpensive to add new columns and it is done on a row-by-row basis. Column-oriented stores like HBase are a good fit for scenarios where only a partial scan of a table is required, and no query search through an entire table data. This should be avoided in cases where a frequent full table scan is needed, such as an average or a summation.

Now you might be asking why you need HBase when you already have relational databases. Relational databases have traditionally had the single-instance model, although you do have distributed databases, but a limiting factor of this is the amount that they can scale. The throughput that you get from HBase is typically higher compared to what you get from stand-alone or distributed databases. Most of these installations typically do not provide beyond 8 or 16 nodes. HBase offers thousands of nodes. And it makes petabytes of data instantly available to you.

You can have any number of columns in an HBase table, but you have to define the column family first. You can define the column family at table creation time or after you created the table. Columns are added to the column family when data is inserted into it.

HBase exhibits fault tolerance by using data replication provided by HDFS. In HDInsight, this is handled by Azure Blob storage. HBase is fault tolerant because its writes are atomic and consistent, which means that HBase guarantees saving data to disk, no matter what situation arises. It has automatic failover, so if a region goes down, another one takes over the responsibility. Also, it distributes read/write loads using automatic sharding and load balancing of tables.

Where to Use HBase?

Basically, HBase was built on Bigtable, which was created for web search. Search engines build indexes that map terms to the webpages that contain them. But there are many other use cases suitable for HBase. A few of them are outlined next.

- **Sensor and IoT data**: High-volume data streams coming from sensors and IoT devices are easily stored using HBase. Data is collected incrementally from various sources. This includes social analytics, time series data, an interactive dashboard, trends, and so forth.

- **Server or website log**: HBase is ideal for storing log data because it allows a different number of columns for different log entries, making it cost-effective in terms of storage and maintenance.

- **Real-time queries**: Phoenix provides a SQL query engine on top of HBase. It can be accessed by JDBC (Java Database Connectivity) or ODBC (Open Database Connectivity) drivers. It enables querying and managing HBase tables by using SQL.

- **Key-value store**: HBase can be used as a key-value store. It is suitable for managing message systems. Facebook uses HBase for its messaging system. It is ideal for storing and managing Internet communications.

■ **Note**   HBase is used by many companies. Facebook uses it for a messaging platform, Twitter uses it for people searches, HubSpot stores primary customer data, and many others are using it for mission-critical operations. Read more about who uses HBase at `http://hbase.apache.org/poweredbyhbase.html`.

The Architecture of HBase

Apache HBase is composed of three types of servers: HBase Master server, the region servers, and ZooKeeper. Clients directly communicate with the region server to read or write data. HBase Master is a lightweight process that handles the assignment of regions and DDL (create, alter, delete, etc.) operations. ZooKeeper provides distributed synchronization and centralized monitoring servers. Figure 6-1 shows the overall architecture of HBase.

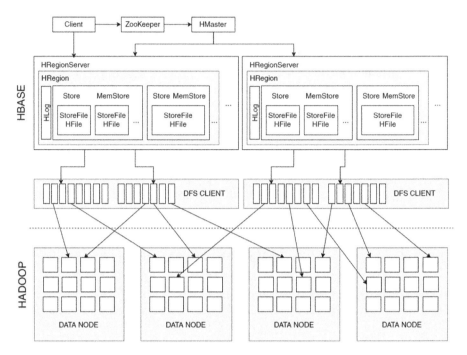

Figure 6-1. *HBase architecture*

To understand HBase architecture, let's assume that there is a large table with simply a key and value pair. You sort this table by the key and then try to spread it across several machines. To do so, you split it into chunks of data. These chunks are called *regions* in HBase. And since this data needs to be served back, entity that does this is a *region server*. A region server can have multiple regions. The metadata about region mapping is stored in ZooKeeper.

When writing data to HBase, it guarantees write consistency and that utilizes HLog. Store files are the representation of the HBase file on disk. And every write from the user goes to HLog, as well as to MemStore. The DFS client is HDFS for on-premise installations; whereas on HDInsight on Azure, it is Azure Blob storage.

HBase HMaster

As you saw earlier, HBase HMaster handles region assignment and DDL (create, update and delete tables) operations. There are two main responsibilities of HMaster server, as described in the following.

- Region monitoring, which includes assigning and reassigning regions to the region server, listening for notification from ZooKeeper for failed regions, and load balancing.

- Creating, altering, and deleting tables issued from client. Typically handles all DDL operations.

HRegion and HRegion Server

HBase divides data into smaller chunks and calls them regions. A region contains all of a table's data, from the start row key to the end row key, including all rows and columns for those row keys. Regions are hosted on region servers, which are nodes of an HBase cluster. A region server can have thousands of regions. The following are the components of HRegionServer.

- HLog: A write ahead log implementation that stores all the edits to the HStore. It stores data that hasn't yet been pushed to permanent storage. It is used in case of failure for recovery. It is a physical file that is not in memory.

- BlockCache or StoreCache: Serves as the cache for recently read data. When full, removes data in FIFO (first in, first out) order.

- MemStore: Stores sorted data that is yet to be written to disk. There is one MemStore per column family per region. Once MemStore reaches a certain size, it is flushed to disk.

- HFile: Stores actual data on a file in sorted in key-value pair order. You can go into your cluster and on HDFS/Azure Blob storage to see these files.

Initially, there is only one region per table (if split; not specified at table creation). Once a table accumulates enough data, HBase splits it into smaller regions, making two regions, with half the original data in each new region. At the time of the split, both regions reside on the same region server. Once HMaster is aware of the region split, it moves them to another region server as required per load balancing. If everything is balanced, then there is no movement. This movement won't serve data from the region server; instead, until major compaction happens, data continues to reside on the remote server. It is major compaction's job to move data to the region server's local node (major compaction is discussed later in the chapter).

The region split policy can be configured. The default region split policy for HBase 0.94 and lower splits the regions when the total data size for one of the stores (corresponding to a column family) in the region gets bigger than the default value, which is 10GB. Since HBase version 0.94, the default split policy has been called the upper bound region split policy. In this policy, split size is increased based on data divided to all regions. The following is the formula for the maximum store file size.

```
minimum of (R^2 * MemStore flush size, max store size)
```

For example, the default MemStore flush size is 128MB, that max store size is 10GB, and then the first split happens at 128MB. After that, it increases as regions increase. The next split happens at 512MB (2^2 * 128), 1152MB (3^2 * 128), 2GB (4^2 * 128), and so on. After the ninth split, the size becomes bigger than 10GB, and from there, splits happen at 10GB.

ZooKeeper

ZooKeeper is a high-performance coordination service for distributed applications like HBase. HBase relies on ZooKeeper for cluster configuration and management. It coordinates, communicates, and shares the state between the HMaster and HRegionServer. Region servers and masters send ZooKeeper heartbeat signals to tell that they are alive. In case of failure, ZooKeeper notifies interested parties to take action; for example, for a master failure, it notifies the standby/inactive master about it.

HBase Meta Table

A meta table is an HBase table that keeps a list of regions and their start row key information. In Azure HDInsight, this is stored in Azure Blob storage, which means that if you delete your cluster and re-create it using the same Blob storage, then you get all the tables back as is. A meta table stores the key and the value, where the key consists of a region start row key and region id, and the value contains the region server. ZooKeeper stores information about where to find the meta table.

Read and Write to an HBase Cluster

When a read request comes to an HBase cluster, the following steps are performed. They are also illustrated in Figure 6-2.

1. The client request comes, which requires metainformation from ZooKeeper.

2. A query is executed to find a region server that holds the data corresponding to the row key that the client wants to fetch.

3. This metainformation is cached for later use by client.

4. Once the region server is located, the client goes to the region server and fetches the row key related data, either from MemStore or from HFile.

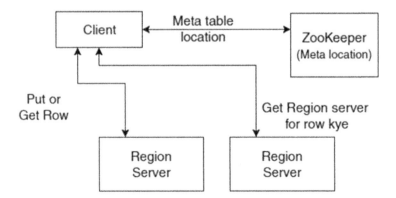

Figure 6-2. *Read data flow in HBase cluster*

When the client wants to write data to an HBase table, following steps are performed. They are also illustrated in Figure 6-3.

1. A Put request is generated by the client.

2. It finds the meta details of the region server that handle the write.

3. Data is handed over to HLog, which stores it physically on disk until it is written to HFile. Before writing data to HFile, if failure occurs, HLog data is used to recover from failure. HLog data is used. HBase makes sure it always writes it to disk.

4. Once data gets through to HLog, it is transferred to MemStore as well, which maintains a sorted list and flushes it out to disk when full.

5. Once data is written to MemStore, acknowledgment is sent for a put request.

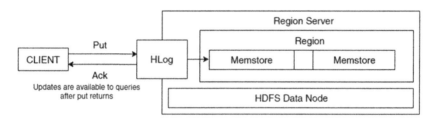

Figure 6-3. *Write a data flow in HBase cluster*

Once MemStore is full, it flushes out the data to disk. Each flush creates a new HFile, which is compacted later if it is small. MemStores are maintained per column family, which means if any of the column families becomes full, then all MemStores are flushed together. This is one of the reasons that you should keep your column families to minimum. HBase supports approximately 10 column families per table (`http://hbase.apache.org/book.html#schema`).

After the first read, data is available at a block cache and HFile. Also, updates to the same data can cause another copy of the data in MemStore. Now the same data with different/same versions can reside in three different places. So, in subsequent reads, HBase first looks into block cache for the row cells that were cached recently. If it cannot find it in block cache, then next it looks in MemStore, which contains recent writes. If the scanner cannot find all the rows, then it loads rows and cells from HFile. Also, there can be multiple versions of the row; hence, it has to load multiple HFiles, which causes performance hits; this phenomenon is called *read amplification*. HBase overcomes this issue using compaction, which you see later in the chapter.

HFile

Actual data is stored in an HFile, which contains sorted key-value pairs. MemStore creates an HFile when it is full. Every flush creates a new HFile. As it is a sequential write and tries to avoid moving the disk driver head, it is very fast. It has multi-layer indexes; hence, data seek is really fast and doesn't require a whole file scan. The HFile index is like a B-tree (binary tree), giving it efficient and fast lookup capability. When it is required to read an HFile, the relevant index is loaded into memory and lookup is performed on it. Finally, with a single disk seek, data is read from the file.

Major and Minor Compaction

When MemStore gets flushed to disk, it creates small files. As I have already discussed, this is not good for the performance of the HBase cluster. Hence, HBase automatically picks small HFiles and combines them to fewer but larger files. This automatic combining of files in HBase is called *minor compaction*. After a merge is complete, the small HFiles are deleted from the system.

On the other hand, major compaction merges and rewrites all the HFiles in a region to one HFile per column family. Deleting in HBase is not a physical deletion, rather a row is marked for deletion; hence causing duplicate keys to become scattered around multiple files. Major compaction removes these duplicate/deleted keys and merges the files together. Also, in the case of a region server failure and region split, there are some region movements required to balance the load. This process also happens in major compaction. It significantly improves the read performance because fewer files need to be accessed. Since this moves a lot of files and data, it involves a lot of disk and I/O operations. This might lead to high network traffic, which is a phenomenon called *write amplification*.

Creating an HBase Cluster

So far, you have only created Hadoop-based clusters. Azure HDInsight provides many different cluster types, as discussed in Chapter 2. To work with HBase, you need master nodes, ZooKeeper nodes, and worker nodes. HDInsight provides an easy-to-deploy HBase cluster type that automatically creates all of these nodes. There are multiple ways to create an HBase cluster—the Azure portal, PowerShell, and Azure Resource Manager templates. You already know how to create a cluster using the Azure portal and PowerShell. You just need to change the cluster type to HBase, as discussed in Chapter 2. In this section, you explore provisioning resources in Azure through templates.

The Azure Resource Manager (ARM) template is a JSON file with all the different resources and configuration options to deploy/provision/configure resources in Azure. You will use the template available at `https://hditutorialdata.blob.core.windows.NET/armtemplates/create-linux-based-hbase-cluster-in-hdinsight.json`. This template creates a Hadoop cluster with two HMaster nodes, three ZooKeeper nodes, and two (default value) worker nodes. If you don't need to create such a JSON template, and you only use them to deploy the resources, then you don't need to worry about understanding the content of the template.

■ **Note**   There are many more ARM templates provided and maintained by Microsoft at
`https://github.com/Azure/azure-quickstart-templates`.

The following procedure creates a Linux-based HBase cluster.

1. Click the following URI, which takes an ARM template URL
    and opens an easy-to-configure options page to create an
    HBase cluster.

    `https://portal.azure.com/#create/Microsoft.Template/`
    `uri/https%3A%2F%2Fhditutorialdata.blob.core.windows.`
    `NET%2Farmtemplates%2Fcreate-linux-based-hbase-cluster-in-`
    `hdinsight.json`

2. Once the template loads, it opens a Custom Deployment
    blade, where you can enter following details.

    •   **Subscription**: Select the Azure subscription to be utilized by
        your cluster.

    •   **Resource group**: You can choose to create a new resource
        group or select an existing group.

    •   **Location**: Specify where your cluster and data will reside,
        such as East US, North Europe, Southeast Asia, and so forth.

    •   **ClusterName**: The cluster's unique name to access it, such as
        `https://{clustername}.azurehdinsight.NET`.

    •   **Cluster credentials**: The cluster username and password to
        open the dashboard. These credentials are used to submit
        jobs and get into the cluster. The default username is admin.

    •   **SSH credentials**: The username and password to get onto a
        Linux machine.

    •   **Cluster Worker Node count**: The number of worker nodes
        that you want the cluster to have. If you are creating a cluster
        for learning purposes, then two nodes are enough.

3. The last steps are to agree to the terms and conditions, and
    then click the Purchase button.

It takes time to provision all the nodes. Once the process is completed (a notification
on Azure portal states this), you can open the HBase dashboard. Enter the URL (e.g.,
`https://{clustername}.azurehdinsight.NET`) and click the HBase node from the
services list. At the top of the HBase service summary page, select the active node in the
Quick Links menu, and then click the HMaster UI menu item. Now you can see various
stats about HBase region servers, tables, Standby Master, and so forth as shown in
Figure 6-4.

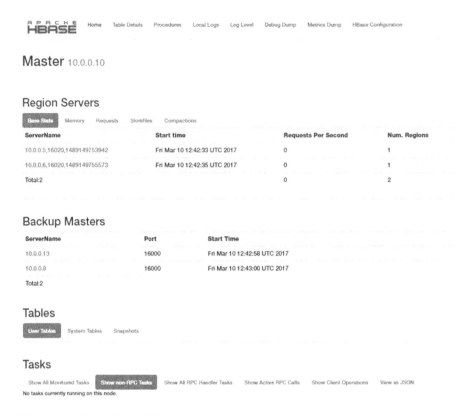

Figure 6-4. HBase Master UI

You have gone through HBase and seen how it works. Now it is time to dive into the code and learn how to work with HBase using data.

Working with HBase

HBase provides a shell to execute commands, insert data, search, and retrieve data. You can also use .NET SDK to read/write data in HBase. You will look at both of these options in the coming sections. The HBase shell is an easy-to-use command-line utility, so let's start with that first.

HBase Shell

To open HBase shell, you need to SSH to your Linux cluster (for a Windows cluster, you can RDP to it and use the Hadoop command-line utility). You use PuTTY to open the SSH connection. Once the connection is made, execute the HBaseshell command to get into the HBase shell. Once you are in the shell, you can execute a few commands to verify that everything is up and running correctly. Execute the status 'summary' and status 'simple' commands to see what is available. The result of these commands is shown in Figure 6-5.

```
sshuser@hn0-hdi1: ~                                    _  □  ×

hbase(main):002:0> status 'summary'
1 active master, 2 backup masters, 2 servers, 0 dead, 1.5000 average load

hbase(main):003:0> status 'simple'
active master:  10.0.0.10:16000 1485418142680
2 backup masters
    10.0.0.5:16000 1485418144075
    10.0.0.9:16000 1485418143660
2 live servers
    10.0.0.13:16020 1485418152705
        requestsPerSecond=0.0, numberOfOnlineRegions=1, usedHeapMB=261, maxHeapM
B=3948, numberOfStores=1, numberOfStorefiles=1, storefileUncompressedSizeMB=0, s
torefileSizeMB=0, memstoreSizeMB=0, storefileIndexSizeMB=0, readRequestsCount=0,
 writeRequestsCount=0, rootIndexSizeKB=0, totalStaticIndexSizeKB=0, totalStaticB
loomSizeKB=0, totalCompactingKVs=0, currentCompactedKVs=0, compactionProgressPct
=NaN, coprocessors=[SecureBulkLoadEndpoint]
    10.0.0.11:16020 1485418141093
        requestsPerSecond=0.0, numberOfOnlineRegions=2, usedHeapMB=411, maxHeapM
B=3948, numberOfStores=2, numberOfStorefiles=2, storefileUncompressedSizeMB=0, s
torefileSizeMB=0, memstoreSizeMB=28, storefileIndexSizeMB=0, readRequestsCount=5
1, writeRequestsCount=32673, rootIndexSizeKB=0, totalStaticIndexSizeKB=0, totalS
taticBloomSizeKB=0, totalCompactingKVs=0, currentCompactedKVs=0, compactionProgr
essPct=NaN, coprocessors=[MultiRowMutationEndpoint, SecureBulkLoadEndpoint]
0 dead servers
Aggregate load: 0, regions: 3

hbase(main):004:0>
```

Figure 6-5. *HBase cluster status*

In Figure 6-5, you can see that there is one active master and its network location, two backup masters, and two worker nodes. Also, there are other details about the worker nodes, such as requests per seconds, regions, and so forth.

Create Tables and Insert Data

HBase stores data in rows and columns, which are inside a table. You can relate a table in HBase to that of a traditional relational database table. But columns are quite a different concept here. As already discussed, columns are created when data is inserted and the column should be included in one of the column families. At table creation time, you only specify the column family. So, let's look at how to store simple employee contact information.

```
Name: Joe Hayes
Email: joe@contoso.com, joe@live.com, Phone: 230-555-0191, 646-555-0113,
508-555-0163
```

The first thing to do is divide data into column families. Table 6-1 shows one of the ways that you can divide data into column families in a table structure.

Table 6-1. *Employee Table*

	Column Family: name		Column Family: email		Column Family: phone		
ID	firstname	lastname	work	personal	home	office	mobile
E1	Joe	Hayes	joe@contoso.com	joe@live.com	230-555-0191	646-555-0113	646-555-0113

It is difficult to grasp the column family concept at first. But assume that some employees have two personal email addresses or only one mobile phone as the contact number. In such cases, a column family can have more/less columns than the previous record. Unlike a traditional relational database, you can create columns when inserting data. Hence, each record can have a different number of columns, optimizing both storage and performance. The following is the procedure to create a table and insert data into it.

1. The create command creates a new table. Name it employee and create three column families in it.

    ```
    create 'employee', 'name', 'email', 'phone'
    ```

■ **Note** HBase is case sensitive. A table named 'employee' is not same as 'Employee'. The same applies to a column and a column family.

---

2. After the table is created, you can view existing tables by using the list command.

3. To insert data, the simplest way is to use the put command. Syntax for the put command is put 'table', 'rowkey', 'columnfamily:column', 'value'. Based on this syntax, the code to insert all data for an employee table looks like the following.

    ```
    put 'employee', 'E1', 'name:firstname', 'Joe'
    put 'employee', 'E1', 'name:lastname', 'Hayes'
    put 'employee', 'E1', 'email:work', 'joe@contoso.com'
    put 'employee', 'E1', 'email:personal', 'joe@live.com'
    put 'employee', 'E1', 'phone:home', '230-555-0191'
    put 'employee', 'E1', 'phone:work', '646-555-0113'
    put 'employee', 'E1', 'phone:mobile', '646-555-0113'
    ```

4. After all the rows are inserted, you can view data using the scan command shown next. The result shown in Figure 6-6.

    ```
    scan 'employee'
    ```

In Figure 6-6, you can see that for a single row, the command returns multiple records, one for each column. Also, note that there is a timestamp field along with the column value. This timestamp is used to return the appropriate versions of flashback queries.

```
sshuser@hn0-hdi1: ~                                    _  □  ×

hbase(main):018:0> scan 'employee'
ROW                   COLUMN+CELL
 E1                    column=email:personal, timestamp=1485428512429, value=joe@live.com
 E1                    column=email:work, timestamp=1485428512391, value=joe@contoso.com
 E1                    column=name:firstname, timestamp=1485428512197, value=Joe
 E1                    column=name:lsstname, timestamp=1485428512351, value=Hayes
 E1                    column=phone:home, timestamp=1485428512472, value=230-555-0191
 E1                    column=phone:mobile, timestamp=1485428523584, value=646-555-0113
 E1                    column=phone:work, timestamp=1485428512528, value=646-555-0113
1 row(s) in 0.1050 seconds

hbase(main):019:0>
```

Figure 6-6. *Employee table row*

HBase Shell Commands

There is a long list of commands that you can execute on an HBase shell. This section looks at a few of the very common commands.

- The count command returns the number of rows in a table. By default, it returns a count every 1000 rows. You can change this default behavior by specifying an interval to the count command.

  ```
  hbase> count 'table'
  hbase> count 'table', INTERVAL => 10000
  ```

- The get command returns row or cell contents. It takes a table name and row key as the minimum parameters. Optionally, you can specify a dictionary of column(s), timestamp, time range, and versions.

  ```
  hbase> get 'table', 'rowkey'
  hbase> get 'table', 'rowkey', {COLUMN => 'column1', TIMESTAMP => ts1}
  hbase> get 'table', 'rowkey', {TIMERANGE => [ts1, ts2]}
  hbase> get 'table', 'rowkey', {VERSION => 4}
  ```

- The scan command, as the name suggests, scans a table. It takes a table name and optionally many different scanner specifications. Scanner specifications include TIMERANGE, FILTER, LIMIT, STARTROW, ENDROW, TIMESTAMP, MAXLENGTH, COLUMNS, and so forth. FILTER is discussed later in the chapter.

  ```
  hbase> scan 'table', {COLUMNS => 'c:colname', LIMIT => 10}
  hbase> scan 'table', {FILTER => "(PrefixFilter ('row')) AND
  (TimestampsFilter (123, 456))"}
  ```

- The disable command takes a table name and starts disabling the table. Disabling a table does not allow the user to query it or insert data. It is still available through the list and exists commands.

- The drop command removes the table from HBase, taking the table name only. Please note that before dropping the table, it must be disabled.

- The delete command puts a delete cell value at a specified table/row/column and optionally, timestamp coordinates it.

- The deleteall command deletes all cells in a given row; pass a table name, row, and optionally a column and timestamp.

The scan command has a filter as one of its options. Filters are useful when it comes to querying data out of HBase. A filter language was introduced in Apache HBase 0.92. The following are the available filters.

- KeyOnlyFilter: The simplest filter, it takes no parameter and returns the key component of a row.

- FirstKeyOnlyFilter: A filter that only returns the first key value from each row. This filter can be used more effectively to perform a row count operation.

- PrefixFilter: Takes one argument, a prefix of a row key, and returns all the rows that match the prefix with the row key prefix.

- ColumnPrefixFilter: Similar to PrefixFilter, but in a column. It takes one argument, a column prefix, and returns only those key-values present in a column that starts with the specified column prefix.

- PageFilter: Limits the results to a specific page size. It terminates scanning once the number of rows satisfying the filter is greater than the given page size.

- QualifierFilter: Takes two parameters: an operator like equal, greater than, not equal, and so forth, and a byte array comparator for the column qualifier portion of a key.

- ValueFilter: Filters based on column value. It takes a comparator and an operator as parameters.

Using .NET SDK to read/write Data

Microsoft provides .NET SDK to work with HDInsight clusters, which you used in previous chapters. For HBase, there is the Nuget library to directly issue commands from .NET code. This allows you to seamlessly integrate HBase into .NET applications. To demonstrate the usage of .NET SDK for HBase, let's implement the following scenario.

You need to show the user all the places that fall under a specific postal code. There can be one or more places for a single postal code.

For this, you will use the data from `http://geonames.org` and download Indian postal codes from `http://download.geonames.org/export/zip/IN.zip`. I've changed the data a bit to form a row key as per the filter need you have. The row key contains the state, city, community, and postal code combination. You group data by postal code and flatten the group into a single row, with a row key as the first column. The rest of the columns are location names (multiple columns). You can download it from `http://bit.ly/indian-post-codes`.

One of the important parts of an HBase column is that you do not need to remember column names. In our case, you want to store the postal code and all the places with the same postal code in the same row. Some postal codes only contain one place, while some may contain ten places. This is an ideal scenario for an HBase column family. HBase is optimized for this. The following is the sample data from the `IN.tsv` file (`http://bit.ly/indian-post-codes`)—handpicked and trimmed for brevity—to show what the data looks like.

```
IN|Delhi|Central Delhi|New Delhi|110004          Rashtrapati Bhawan
IN|Maharashtra|Mumbai|Mumbai|400001              Stock Exchange
                                                 Town Hall (Mumbai)
IN|Karnataka|Bangalore|Bangalore North|560001    Mahatma Gandhi Road
                                                 Highcourt
IN|West Bengal|Kolkata|Kolkata|700001            Treasury Building
                                                 R.N. Mukherjee Road
```

Each line has two places that share a single postal code. Now let's write the code to insert the data into the HBase table.

Writing Data

To write data from the local file, follow these steps.

1. Create a C# console application in Visual Studio 2012 or higher by clicking New ➤ Project from the File menu and then selecting Visual C# ➤ Windows ➤ Console Application.

2. Add a Nuget reference to the Microsoft.HBase.Client. This is a REST client for HBase. Installing this Nuget will add dependencies as well, which includes protobuf-net and the fault-handling library.

3. Read the data and create an in-memory list of row keys (string) and places (array of string). Assume that there is a IN.tsv file in the same folder as the application and read the content. The following method should be placed inside the `Program` class alongside the `Main` method. It will be called from your `Main` method directly; hence, it is a static method.

```
private static List<Tuple<string, string[]>> ReadData()
{
    // Load data from CSV line by line
    var reader = new System.IO.StreamReader(System.IO.File.OpenRead(@"IN.tsv"));
    string line;
    List<Tuple<string, string[]>> postCodes = new List<Tuple<string, string[]>>();

    while (!reader.EndOfStream)
    {
        line = reader.ReadLine();

        string rowKey = line.Substring(0, line.IndexOf('\t'));
        string[] places = line.Substring(line.IndexOf('\t'))
            .Split(new char[] { '\t' }, StringSplitOptions.RemoveEmptyEntries);

        postCodes.Add(new Tuple<string, string[]>(rowKey, places));
    }
    Console.WriteLine($"Post code read completed with total: {postCodes.Count}");
    return postCodes;
}
```

> In the ReadData method, you read a file line by line. Each
> line is separated by a tab character, which makes the first
> element the row key and the rest of the string are places, again
> separated by a tab character.

4. Write the data in to HBase. For this, create a new HBaseWriter
   class. Check whether you have the table in HBase or not. If
   you don't, then create it. You will write in a batch of 1000
   records to optimize the write speed. The following is a method
   called WriteData the writes data into an HBase cluster. The
   following is the code for the HBaseWrite class.

```
public class HBaseWriter
{
    // HDinsight HBase cluster and HBase table information
    string _clusterName = "https://{0}.azurehdinsight.NET/";
    string _hadoopUsername;
    string _hadoopPassword;
    const string HBASETABLENAME = "postcodes";

    public HBaseWriter(string clusterName, string username, string password)
    {
        _clusterName = string.Format(_clusterName, clusterName);
        _hadoopUsername = username;
        _hadoopPassword = password;
    }
```

```
    public async Task WriteDataAsync(List<Tuple<string, string[]>> postcodes)
    {
        HBaseClient client;
        ClusterCredentials credentials = new ClusterCredentials(
new Uri(_clusterName), _hadoopUsername, _hadoopPassword);
        client = new HBaseClient(credentials);

        // create the HBase table if it doesn't exist
        if (!client.ListTablesAsync().Result.name.Contains(HBASETABLENAME))
        {
            TableSchema tableSchema = new TableSchema();
            tableSchema.name = HBASETABLENAME;
            tableSchema.columns.Add(new ColumnSchema { name = "p" });
            client.CreateTableAsync(tableSchema).Wait();
            Console.WriteLine("Table \"{0}\" is created.", HBASETABLENAME);
        }
        int rows = 0;
        CellSet set = null;
        int pageSize = 1000;

        foreach (Tuple<string, string[]> data in postcodes)
        {
            if (rows % pageSize == 0)
                set = new CellSet();

            // Create a row
            var row = new CellSet.Row { key = Encoding.UTF8.GetBytes(data.
            Item1.Substring(3)) };

            foreach (string place in data.Item2)
            {
                // Add columns to the row
                var value = new Cell
                {
                    column = Encoding.UTF8.GetBytes("p:" +
                        Guid.NewGuid().ToString().Substring(0, 6)),
                    data = Encoding.UTF8.GetBytes(place)
                };
                row.values.Add(value);
            }

            // Add row to CellSet
            set.rows.Add(row);

            // Write the postal codes and places to the HBase table
            rows++;
            if (rows % pageSize == 0 || postcodes.Count == rows)
            {
```

```
            await client.StoreCellsAsync(HBASETABLENAME, set);
            Console.WriteLine("\tRows written: {0}", rows);
        }
    }
  }
}
```

The important thing to notice here is the column name. You are never going to fetch the data based on column names. To us, all places are the same. Hence, you just put a random six-character string, which you can see in the code after the Add columns to the row comment.

Reading/Querying Data

After data is written to HBase, it is time to learn how to query it back. Let's use the same project from the previous section and add a new class to it named HBaseReader. You'll read the postal codes and the state-city-community (e.g., Delhi-Central Delhi-New Delhi). You can execute a scan command to query from an HBase shell. Here you don't have a full row key; hence, you use PrefixFilter with the scan command.

```
scan 'postcodes', {FILTER => "PrefixFilter('Delhi|Central Delhi|New Delhi')" }
```

It returns 11 rows, starting with the given prefix filter. Now, let's see how this is done in C#. In C#, you need to create a scanner that takes parameters, like the prefix filter and batch size, to return. Once you create a scanner object, you can call HBaseClient's ScannerGetNextAsync method to fetch matching records. Since it returns data in batches, you need to loop on a batch of data until you get all the records out. the complete HBaseReader class is shown next.

```
public class HBaseReader
{
    // HDinsight HBase cluster and HBase table information
    string _clusterName = "https://{0}.azurehdinsight.NET/";
    string _hadoopUsername;
    string _hadoopPassword;

    const string HBASETABLENAME = "postcodes";

    public HBaseReader(string clusterName, string username, string password)
    {
        _clusterName = string.Format(_clusterName, clusterName);
        _hadoopUsername = username;
        _hadoopPassword = password;
    }
```

```
public async Task QueryDataAsync(string state, string city, string community)
{
    ClusterCredentials creds = new ClusterCredentials(
        new Uri(_clusterName), _hadoopUsername, _hadoopPassword);
    HBaseClient client = new HBaseClient(creds);

    string startRow = $"{state}|{city}|{community}";
    Scanner scanSettings = new Scanner()
    {
        batch = 1000,
        startRow = Encoding.UTF8.GetBytes(startRow),
        filter = new PrefixFilter(Encoding.UTF8.GetBytes(startRow)).
        ToEncodedString()
    };

    // Make async scan call
    ScannerInformation scannerInfo =
        await client.CreateScannerAsync(HBASETABLENAME, scanSettings,
            RequestOptions.GetDefaultOptions());

    CellSet next;

    while ((next = await client.ScannerGetNextAsync(
        scannerInfo, RequestOptions.GetDefaultOptions())) != null)
    {
        foreach (CellSet.Row row in next.rows)
        {
            var places = row.values
                .Where(o => Encoding.UTF8.GetString(o.column).
                StartsWith("p:"));
            Console.WriteLine(Encoding.UTF8.GetString(row.key));
            foreach (var item in places)
            {
                Console.WriteLine("\t" + Encoding.UTF8.GetString(item.data));
            }
            Console.WriteLine("");
        }
    }
}
```

And finally, the main method looks like following.

```
static void Main(string[] args)
{
    // Read data from file and create in memeory list
    List<Tuple<string, string[]>> postCodes = ReadData();

    // Save it to HBase
    HBaseWriter writer = new HBaseWriter("{ClusterName}", "{Username}",
    "{Password}");
    writer.WriteDataAsync(postCodes).Wait();

    // Query data back
    HBaseReader reader = new HBaseReader("{ClusterName}", "{Username}",
    "{Password}");
    reader.QueryDataAsync("Delhi", "Central Delhi", "New Delhi").Wait();

    Console.ReadKey();
}
```

With this code, you can save and read data from HBase.

Summary

In this chapter, you explored HBase and learned how easy it is to query massive data with it. Apache HBase is a NoSQL database on top of Hadoop that has fast read/write speed with consistency. HBase guarantees that it will write everything to disk. The region and region server concepts provide high flexibility in reading and writing data. An automatic sharding process takes the load off developers, allowing them to focus on implementing business scenarios. It can easily be configured from the HBase dashboard. HBase Shell provides a powerful command-line tool to interact with HBase when developing applications and performing quick processing. HBase .NET SDK gives developers the ability to integrate HBase read/write seamlessly into their own applications.

In the next chapter, you look at Apache Storm, a real-time stream-processing system.

CHAPTER 7

■ ■ ■

Real-Time Analytics with Storm

So far, you've seen how to work with batch data processing in Hadoop. Batch processing is used with data at rest. You typically generate a report at the end of the day. MapReduce, Hive, and HBase all help in implementing batch processing tasks. But there is another kind of data, which is in constant motion, called *streams*. To process such data, you need a real-time processing engine. A constant stream of click data for a campaign, user activity data, server logs, IoT, and sensor data—in all of these scenarios, data is constantly coming in and you need to process them in real time, perhaps within a window of time. Apache Storm is very well suited for real-time stream analytics. Storm is a distributed, fault-tolerant, open source computation system that processes data in real time and works on top of Hadoop.

This chapter looks into Storm on HDInsight to understand its components and workings. You will submit a Storm topology to process real-world streams of data.

Overview

Apache Storm is a reliable way to process unbounded streams of data. Similar to what Hadoop does for batch processing, Storm does for real-time processing. It is simple, easy to implement, and takes away the complexity of thinking and writing code that can run on multiple nodes simultaneously. I have been talking about streams of data. So what is it? A *stream* is an unbounded sequence of data. *Unbounded* means that it doesn't have a start or an end. Hence, it is not like reading a file or records from a database. You will never get to the end of a stream because it is a never-ending continuous stream of data. Think of Twitter as a stream of data; continues without end.

So, unbounded streams mean that processing cannot wait until all the data is received, because that will never happen. It is not like running a query against a table, as you will never have all the data in a table. This signifies that processing has to be continuous as well. That is why you continuously process and get insight out of a stream as it passes through the data pipeline. And processing cannot be done on everything you have in a stream; for example, aggregating a whole stream is not as useful as one

© Vinit Yadav 2017
V. Yadav, *Processing Big Data with Azure HDInsight*, DOI 10.1007/978-1-4842-2869-2_7

on a temporal window. For example, if you are getting a stream from Twitter, and you are monitoring hashtags matching to a company or a person, then you might be more interested in knowing the most used hashtag in the past hour or past day, compared to the most used hashtag since the beginning of Twitter.

Now that you understand what a data stream is, let's discuss Apache Storm, which can be used to make sense of data streams. Apache Storm is the project built on Hadoop technologies to process streams of data in a reliable and efficient manner. To do so, Storm uses the concept of *topology*. A topology consists of spouts and bolts. A *spout* is the part of topology that is responsible for getting data from an external stream and feeding it into the topology for processing. There can be one or more spouts in a topology. They connect to a source to fetch data. The spout then emits a stream itself. Each emission consists of a *tuple*, which is effectively a record or a row of data.

Bolts are also within a topology. Bolts consume the output stream from spout(s) or from another bolt(s), perform some operation on the existing tuples, and emit it further down the pipeline. Bolts are responsible for aggregating, counting, storing, or doing any other processing on a tuple. A bolt can also emit the same, or different, or multiple streams to the next bolts in the pipeline. This way, you can build any complex topology.

Ultimately, a topology has spouts, which convert a data source into a stream of tuples. Bolts act on tuples; they contain the logic to operate on those tuples. And this runs continuously. You submit a topology once, and it continues to process until you stop it.

To compare real-time analytics with traditional batch analytics, consider Figure 7-1. Traditional processing requires running an analytic query against stored historical data. For example, consider an app that calculates the cycling distance of a user who has cycled every month using global positioning system (GPS) location data. Traditionally, in batch processing mode, the app calculates the total distance at the end of each month, based on the data transmitted by the user over the previous month, which is stored in a database. In contrast, stream processing performs continuous analysis to keep running totals that are updated moment by moment as GPS data comes in while the user is actually cycling.

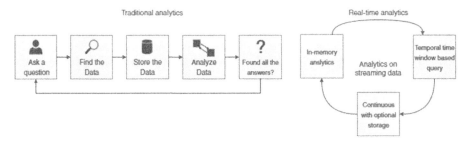

Figure 7-1. *Traditional vs. real-time analytics*

■ **Note**   Storm is designed by Nathan Marz and open sourced by Twitter.

In the preceding case, a traditional approach answers questions by using historic data, while in the second case, answers are continuously updated using streaming data. Streaming analytics is different from a simple traditional database because traditional analytics load all the data first; even if it is in memory, all the data has to be loaded before running a query. On the other hand, real-time analytics continuously monitors data that is modified over time. For example, GPS data can be continuously monitored to identify if the user is deviating from his usual path or to motivate him to go further once he reaches his maximum distance.

Now that you have basic idea about Storm, it is a good time to discuss its benefits:

- **Ease of use**: Apache Storm is built on Hadoop; hence, as a developer, you don't have to worry about the complexity that comes with big data, distributed computing, and scaling. Storm handles the scale-out complexities and the way that code is parallelized, so that you can focus on business logic.

- **Low latency**: Real-time streams can get really fast—like millions of events per second. If you have latency in processing in such a scenario, then your data starts to pile up, and sooner or later, you start losing messages. Things get worse in distributed processing. Storm has been battle-tested by many companies in the real world by running production workloads. Benchmarks have shown that Storm processes more than a million messages per second per node.

- **Scalable**: Just like Hadoop, Storm can be scaled anywhere from a single node to hundreds of nodes.

- **Fault tolerant**: Storm runs on top of Hadoop in a distributed environment, which means that it is run on commodity hardware, which can fail. But Storm guarantees message processing at least one. If any node goes down, then all the messages are automatically replayed to different nodes in the cluster.

- **Reliable**: Let's say that there are many messages coming into Storm, and if any of them fails to complete the whole processing, then Storm knows this and replays that message. On top of that, you can be sure that your message is processed exactly once.

Now that you know the benefits of Apache Storm, I'll list a few scenarios where real-time analytics can be used.

- Internet of Things (IoT)
- Financial transaction and fraud detection
- Network and server log monitoring
- Intelligent traffic management
- Energy grid monitoring
- Social analysis
- Telecom customer churn prediction

145

Storm Topology

As discussed, the Storm topology contains spouts and bolts. A spout emits tuples by consuming from a streaming data source, which can be Twitter, financial data, or data coming from a network port. Normally, Storm is used along with a queuing service like Azure Event Hub or Apache Kafka. Data comes to these queuing services and it is then picked up by a Storm spout. In a topology, data can come from multiple sources. Be it a single source or multiple sources, data comes into Storm only through a spout.

Spouts emit tuples, which you can think of as a record or a row of data in a stream. It is not necessary to emit only one stream; a spout can consume a stream, split it, and emit multiple streams based on the business logic and data. Also, a single spout can consume multiple streams and combine them into a single stream. In either case, a spout has to emit at least one stream.

Tuples emitted by a spout go to a bolt or multiple bolts. The bolt processes it and then emits a tuple for other bolts to process. Say, for example, data coming from a spout gets to the first bolt, which might aggregate it, or perhaps it sorts the data or just logs that the data arrived and passes the same or a new tuple downstream for more processing by other bolts. Just like a spout, a bolt can receive data from multiple streams emitted by different spouts. Usually, a bolt can differentiate between different streams and can act accordingly. So, Storm provides a very flexible and adaptable way of constructing a topology, which makes sense for the data that you want to process. Figure 7-2 shows a sample Storm topology, which illustrates the fundamentals of spouts and bolts. This sample topology consists of multiple spouts, and a bolt can receive data from these multiple spouts.

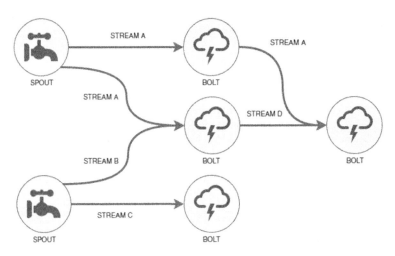

Figure 7-2. Sample Storm topology

Stream Groupings

The next thing to understand in the Storm topology is controlling the flow of tuples. At times, you may want to send the same value tuples to the same bolt so that you can count easily. To control this behavior, there is concept called *stream groupings*. When you define a topology, you build a graph of spouts and bolts. At a more granular level, each bolt is executed as multiple tasks in a topology. And a stream is partitioned among the bolts' tasks. Hence, each task sees subsets of the whole tuple stream. To control how tuples are partitioned and assigned to a bolt, stream grouping is used. Stream grouping is specified on the bolt when you define the topology. The following are a few of the most used stream grouping types.

- **Shuffle grouping**: The most common type of grouping. Shuffle grouping distributes tuples to all bolts in a uniform but random way. Without any particular order, tuples are distributed to all bolts, but every bolt gets an equal number of tuples. Typically, this distributes the load across all the bolts uniformly. So, when you don't have any specific data distribution requirements, use this.

- **Field grouping**: This controls the flow of the tuples to the same bolt based on one or more fields in the tuple. This grouping guarantees that all the same field value tuples will be processed by the same bolt. For example, if you want all the tweets with the same hashtag going to the same bolt for counting, then use the field grouping on the hashtag field of the tweet tuple. Also, partitions happen on the basis of the field value, not based on stream. You can combine multiple streams based on their field value.

- **All grouping**: If you want to send a tuple to all the bolts, then use this grouping. It is not used with data processing. It sends a signal to bolts. For example, to refresh cache data in each bolt every hour, you can send a signal to bolts using all grouping. Also, if you have bolts with filtering that needs to be changed from time to time, then you can use this feature to facilitate changing filters.

- **Global grouping**: You can use this grouping to get all the tuples in a stream to pass through a single bolt. It generally sends the whole stream to the bolt task with the smallest id. You might say this is redundant because you can achieve the same result if you create only a single bolt by defining its parallelism. But what if you only want all the data from a single stream and for the rest of the streams to continue to be processed parallel? In such case, you can achieve your goal by using global grouping. Typically, if you want to carry out some reduce phase in your topology over data coming from the previous step, then global grouping should be used.

- **Direct grouping**: In this grouping, the source decides which bolt receives a tuple. For example, you have a web server log that you want to send to different bolts based on its HTTP response code. Direct grouping can only be used with direct streams.

- **Custom grouping**: This grouping provides a customized processing sequence. This gives maximum flexibility for in-designing topology, based on factors such as type, load, and seasonality.

Storm Architecture

The physical architecture of Apache Storm is also based on a master-slave arrangement, which is similar to the headnode-workernode in Hadoop. In Storm, a master is called Nimbus and a slave is called a *supervisor node*. Also, ZooKeeper coordinates between distributed processes. The following sections discuss the components that make up the Storm architecture.

Nimbus

The master node in Storm is called the Nimbus server. Its job is to distribute application code across worker nodes, determining when new instances are needed, monitoring the running instances for failures, and restarting them as and when needed. Nimbus is partly a task scheduler. If there is no Nimbus server, then no work can be scheduled and no work will be carried out. Hence, Azure HDInsight Storm cluster has two Nimbus servers. One is always in action and other is in passive mode. If active Nimbus goes down, then ZooKeeper as a monitoring service notifies the passive node to become active and takes over the responsibility. Nimbus stores state information in ZooKeeper, so in the event of a failure, the passive node can immediately take charge. And this makes Storm very stable. Also, restarting the Nimbus demon is very fast, which adds more stability.

The supervisor node (discussed in the next section) is a worker node in Storm. It does all the actual work. When a topology deployment request is received, it distributes the code to supervisor nodes, and assigns tasks for spouts and bolts instances to each supervisor. Also, it signals the supervisor nodes to spawn the required worker process for tasks. Afterward, it monitors the status of the tasks assigned to each supervisor node. If it finds any failure, then it reassigns the task to another supervisor. On the other hand, if Nimbus fails and is not restarted, then the topology continues to process because Nimbus doesn't take part in data processing. So, as long as supervisors continue to work without fail, you don't need the Nimbus demon. Once any of the supervisor tasks fails, Nimbus reassigns the task to another node.

Supervisor Node

The slave, or worker, node in Storm is called a *supervisor node*, which actually executes spouts and bolts. Each supervisor node can run multiple instances of each component (spout and bolt). Its primary responsibility includes creating, starting and stopping workers processed to execute assigned tasks. Each worker process has one or more executor, which is essentially just a thread on which an actual component will run. This allows an instance of the same bolt to be created multiple times and distributed across all the nodes in a cluster.

ZooKeeper

When you work in a distributed environment, many processes need to share information, such as configuration settings and other metadata, with each other to coordinate various tasks. To facilitate information sharing in a reliable manner, Storm uses the ZooKeeper service. Apache ZooKeeper is a standalone project that can be used by any application to share information in a distributed environment. Since Storm is also a distributed application, it utilizes a ZooKeeper cluster for the coordination of different processes. ZooKeeper also acts as a communication point between the Nimbus and supervisor process on worker nodes, as illustrated in Figure 7-3.

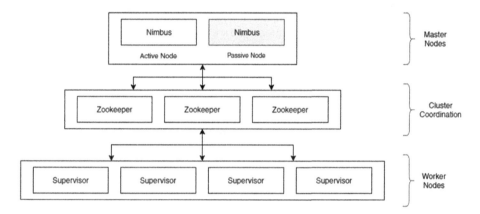

Figure 7-3. *Storm cluster*

Worker, Executor, and Task

Every Storm cluster has one or more worker process. Each worker process has one or more executor. Each executor has one or more tasks, which are nothing but the topology components—spouts and bolts. You can visualize a worker process as shown in Figure 7-4.

149

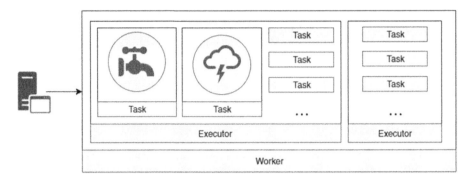

Figure 7-4. Worker, executors, and tasks in supervisor node

Here, the worker process communicates with the Nimbus. And there can be multiple worker processes running across a Storm cluster on your supervisor nodes. Each worker process is isolated from the other by running inside its own Java virtual machine. Each worker process can spawn one or more next level of compute, called an *executor*, which is basically a thread. Inside each executor, there are one or more tasks. A task is a running instance of the Storm component (a spout or a bolt).

Now there are some differences between Java topology and Hybrid .NET topology. As you know, an executor is a thread running in a JVM environment. In a hybrid .NET topology, you can't run .NET code inside a Java process; hence, a .NET component runs its own process. An instance of the SCPHost (discussed in the next section) executable is created per executor, which then runs the component in its own thread. A visual comparison of a Java and a hybrid .NET topology worker process is shown in Figure 7-5.

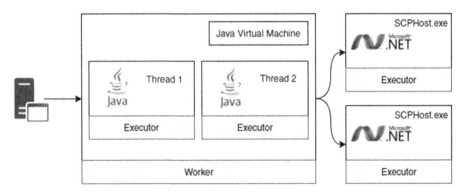

Figure 7-5. Java vs. hybrid .NET topology

This shows that in a hybrid topology, many Java executors are condensed into a few worker processes, but every .NET executor runs in its own separate process. A Java topology has a slight performance benefit because you can route tuples within the worker process and it does not incur the cost of sending a message over the wire. But note that unless you need hundreds of thousands of messages per second on a handful of servers, you can get good enough performance from a hybrid topology.

Creating a Storm Cluster

Azure HDInsight provides many different cluster types, as discussed in Chapter 2. To work with Storm, you need a Nimbus node, a supervisor node, and a ZooKeeper node. HDInsight provides a simple-to-deploy Storm cluster type that automatically creates all of these nodes. There are multiple ways to create a Storm cluster (i.e., Azure portal, PowerShell, and Azure Resource Manager templates). Let's look at the Azure Resource Manager template and the Azure portal to create a Storm cluster.

Using Azure Resource Manager

The Azure Resource Manager (ARM) template is a JSON file with different resource and configuration options to deploy/provision/configure resources in Azure. You will use the template available at `https://hditutorialdata.blob.core.windows.net /armtemplates/create-linux-based-storm-cluster-in-hdinsight.json`.
The following procedure creates a Linux-based Storm cluster.

1. Open the following URI, which takes the ARM template URL and opens an easily configurable options page to create a Storm cluster.

    ```
    https://portal.azure.com/#create/Microsoft.Template/
    uri/https%3A%2F%2Fhditutorialdata.blob.core.windows.
    net%2Farmtemplates%2Fcreate-linux-based-storm-
    cluster-in-hdinsight-35.json
    ```

2. Fill out the following requirements to complete the configuration.

    - **Subscription**: Select the Azure subscription for your cluster from a list of your subscriptions.

    - **Resource group**: Create a new resource group or select an existing group from a list.

    - **Location**: Specify where your cluster and data will reside.

    - **ClusterName**: A unique name for the cluster. Accesses your cluster using `https://{clustername}.azurehdinsight.net`.

    - **Cluster credentials**: The username and password for your cluster to open on the dashboard. These credentials are used to submit jobs and get into a cluster. The default username is admin.

    - **SSH credentials**: The username and password to get onto a Linux machine.

    - **Cluster Worker Node count**: The number of worker nodes you want a cluster to have. If you are creating cluster for learning purposes, then two nodes are enough.

3. Agree to the terms and conditions, and then click the Purchase button.

151

By default, this template provisions two Nimbus nodes and three ZooKeeper nodes, along with whatever number of worker/supervisor nodes that you select. Once you hit the Purchase button, it takes a few minutes to provision the cluster. You can browse the Ambari dashboard by going to `https://{ClusterName}.azurehdinsight.net`.

Using Azure Web Portal

Another way to provision a cluster is to use the Azure portal, which you already learned how to use in Chapter 2. Hence, the following only covers the steps you need to know for a Storm cluster.

- **Cluster type selection**. On the basic configuration settings page, under cluster type selection, choose Storm. You can also choose the operating system (Linux or Windows). Figure 7-6 shows the Windows operating system. In a Linux system, you can choose among different versions of Storm, up to Storm 1.0.1 with HDI3.5. Windows only supports Storm 0.10.0 with HDI 3.3 at the time of writing.

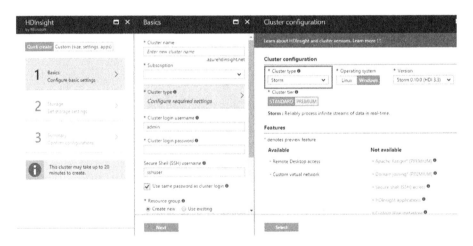

Figure 7-6. *Storm cluster type*

- **Cluster size**. On the custom settings tab, the fourth option is cluster size. Here you can choose the cluster's supervisor nodes count and the Nimbus node, supervisor node, and ZooKeeper node sizes. Figure 7-7 shows four supervisor nodes, with a size D3 v2 virtual machine with four cores each. Both the Nimbus node and the ZooKeeper node are using A3 virtual machine with four cores. There are two Nimbus nodes and three ZooKeeper nodes.

Figure 7-7. Storm cluster size options

Storm UI

The Storm UI (user interface) provides a web-based interface for working with running topologies. It is already included in your HDInsight cluster. Through the Storm UI, you can view the running topology, including its current stats for spouts and bolts.

To open the Storm UI, browse to `http://{ClusterName}.azurehdinsight.net/stormui`, where ClusterName is the name of your cluster. Figure 7-8 shows a sample Storm UI with a topology named Wordcount. Also, you can see topology stats, like tuples emitted and transferred, through topologies in different time windows.

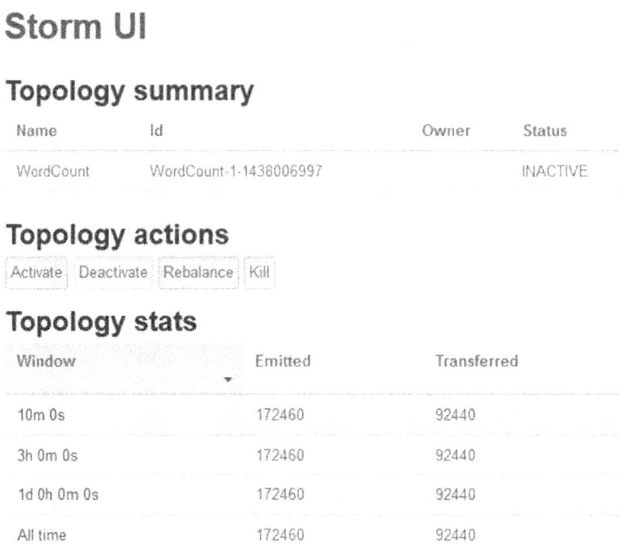

Storm UI

Topology summary

Name	Id	Owner	Status
WordCount	WordCount-1-1438006997		INACTIVE

Topology actions

Activate | Deactivate | Rebalance | Kill

Topology stats

Window	Emitted	Transferred
10m 0s	172460	92440
3h 0m 0s	172460	92440
1d 0h 0m 0s	172460	92440
All time	172460	92440

Figure 7-8. Storm UI

You can view the following information on the Storm UI dashboard.

- **Topology stats**: Information about tuples passed through topology and performance of the same, organized into a time window.

- **Spouts and bolts**: Information about the executors, tasks, and emitted and transferred tuples. Includes spout performance in terms of average latency to process a tuple and acked (acknowledged) or failed messages.

- **Topology configuration**: A summary of the supervisor nodes and the Nimbus nodes, and other system stats.

Other actions are also available.

- **Activate**: Resumes the processing of a deactivated topology.

- **Deactivate**: Pauses a running topology.

- **Rebalance**: When a topology starts, it takes all the nodes into account and tries to get the best use out of them. Once you add or remove nodes from a cluster, the topology may become non-uniform, making a few nodes work more/less than others do. Rebalancing a topology adjusts the parallelism to compensate an increased or a decreased number of nodes in a cluster.

- **Kill**: Permanently terminates the topology and stops all the processing.

In Storm UI, you can drill down to more specific components, such as spouts and bolts, by selecting them from the dashboard page. This opens an information page for the specific component. Figure 7-9 shows the Storm UI's component view for a spout in a topology.

Spout stats

Window ▲	Emitted	Transferred	Complete latency (ms)	Acked	Failed
10m 0s	28500	28500	0.000	0	0
3h 0m 0s	518940	518940	0.000	0	0
1d 0h 0m 0s	562680	562680	0.000	0	0
All time	562680	562680	0.000	0	0

Output stats (All time)

Stream ▲	Emitted	Transferred	Complete latency (ms)	Acked	Failed
default	562680	562680	0	0	0

Figure 7-9. *Storm UI spout stats*

When viewing information about a spout or bolt, selecting the port number in the executor section opens the logs for the specific instance of the component. In our word count sample, a subset of a log might look like following.

```
2017-01-27 14:18:02 b.s.d.task [INFO] Emitting: split default ["with"]
2017-01-27 14:18:02 b.s.d.task [INFO] Emitting: split default ["nature"]
2017-01-27 14:18:02 b.s.d.executor [INFO] Processing received message
source: split:21, stream: default, id: {}, [snow]
2017-01-27 14:18:02 b.s.d.task [INFO] Emitting: count default [snow, 747293]
2017-01-27 14:18:02 b.s.d.executor [INFO] Processing received message
source: split:21, stream: default, id: {}, [white]
2017-01-27 14:18:02 b.s.d.task [INFO] Emitting: count default [white, 747293]
2017-01-27 14:18:02 b.s.d.executor [INFO] Processing received message
source: split:21, stream: default, id: {}, [seven]
2017-01-27 14:18:02 b.s.d.task [INFO] Emitting: count default [seven, 1493957]
```

In this log, you can see the task and executor and their respective work-specific message.

Stream Computing Platform for .NET (SCP.NET)

The stream computing platform (SCP) for .NET or SCP.NET provides .NET C# programmability against Apache Storm on an Azure HDInsight cluster. SCP is a platform to build a real-time, reliable, consistent, and high-performance data processing application. SCP enables .NET developers to run C# code as a spout or bolt while leveraging JVM-based Storm under the cover. The .NET code and JVM communicates over a TCP (Transmission Control Protocol) local socket. As you saw earlier in this chapter, each component (spout and bolt) in .NET has a Java process pair, where the business logic runs in a .NET process as a plugin.

To design a data processing application with SCP.NET, you need to create at least one spout that pulls data from a stream, a bolt to do processing on the output stream of the spout, and a design topology. The topology defines the vertexes and the data flows between the vertexes. To create a spout, bolt, and topology, SCP.NET provides interfaces and methods. Let's have a look at them next.

ISCP-Plugin

This is a base interface for all the plugins. Currently, it is empty, available to provide a common base. It has no functionality for developers to use. The following is the actual interface code.

```
public interface ISCPPlugin
{
}
```

ISCPSpout

ISCPSpout is the interface for non-transactional spout. It comprises three methods, as shown in the following code snippet.

```
public interface ISCPSpout : ISCPPlugin
{
    void NextTuple(Dictionary<string, Object> parms);
    void Ack(long seqId, Dictionary<string, Object> parms);
    void Fail(long seqId, Dictionary<string, Object> parms);
}
```

It is derived from the ISCPPlugin. All the spout implementa tion in your application must implement the ISCPSpout interface. The main method is NextTuple, which is used to emit tuples for downstream components. To emit tuples, there is an Emit in SCP. Context method (discussed later in this chapter). If there is no tuple to emit in NextTuple method, then it should return without emitting anything. And ideally, it should sleep for 10–50 milliseconds, so as not to waste too much CPU.

The Ack and Fail methods are used when ack is enabled in topology (config value nontransactional.ack.enabled is true). The seqId parameter is used to identify a tuple that is acked or failed. And, if ack is not enabled in a non-transactional topology, then the Ack and Fail methods can be left empty. Lastly, the prams input parameter is an empty dictionary that is reserved for future use.

Depending on whether the ack is enabled or not, a different emit method overload will be used.

ISCPBolt

ISCPBolt is the interface for non-transactional bolts. It has only one method, as shown in next code snippet.

```
public interface ISCPBolt : ISCPPlugin
{
    void Execute(SCPTuple tuple);
}
```

The Execute method is called whenever there is a tuple from any of the streams subscribed by the bolt. As you already know, a bolt can subscribe to multiple streams. To identify which tuple came from which stream, there is the GetSourceStreamId in SCPTuple method, which returns the stream name. After identifying the stream, a bolt can go ahead and process the data, save it, and/or emit for the next bolt(s) to process it.

ISCPTxSpout

The ISCPTxSpout interface is available for a transactional spout. And just like a non-transactional spout, it has three methods.

```
public interface ISCPTxSpout : ISCPPlugin
{
    void NextTx(out long seqId, Dictionary<string, Object> parms);
    void Ack(long seqId, Dictionary<string, Object> parms);
    void Fail(long seqId, Dictionary<string, Object> parms);
}
```

The only difference is the NextTx method. It has a seqId out parameter, which is used to identify the transaction. seqId is also used in Ack and Fail methods. Emitted data is stored in ZooKeeper to support replay. When any message fails, it is automatically replayed. Since there is limited capacity in ZooKeeper, a transactional spout should only emit metadata, and not the whole data.

ISCPBatchBolt

ISCPBatchBolt is a transactional bolt interface. It has two methods: Execute and FinishBatch.

```
public interface ISCPBatchBolt : ISCPPlugin
{
    void Execute(SCPTuple tuple);
    void FinishBatch(Dictionary<string, Object> parms);
}
```

Whenever there is a new tuple from any of the streams that a bolt subscribed to, the Execute method is called. And this method is completely similar to a non-transactional bolt. FinishBatch is new method, only available in transactional bolts and only called when a transaction has ended.

SCP Context

SCP.NET provides the Context object, injected into a constructor in a spout and a bolt. The Context object provides a few helpful and required methods to work with Storm. It makes a developer's life easy by providing common-purpose methods and objects. The following describes a few of the important methods and objects.

```
public static ILogger Logger;
public static SCPPluginType pluginType;
public static Config Config { get; set; }
public static TopologyContext TopologyContext { get; set; }
```

- The Logger object provides logging methods like Debug, Error, Info, Warn, Log, and so forth. These methods write to a Storm log, which can be viewed from a specific component.

- pluginType indicates a plugin type's current process (i.e., spout, bolt, TxSpout, BatchBolt, etc.).

- The Config object gives access to the Storm config and plugin config dictionary.

- The TopologyContext class provides information about the component's place within the topology, such as task ids, inputs and outputs, and so forth.

The following are the most-used methods in Context.

- The DeclareComponentSchema method defines the input and output schema of a component (spout and bolt). Once you declare the schema, your code needs to ensure that it emits tuples that obey the schema, or the system will throw a runtime exception.

  ```
  public void DeclareComponentSchema(ComponentStream
  Schema schema);
  ```

- The Emit method sends the tuple to the next component in the topology. An emitted tuple should match the output schema of the component. The output schema is defined by the components only. There are multiple overloads of this method for transactional and non-transactional components. The first overload only takes one argument, which is the actual value to emit. The second overload takes streamId as well, which signifies the stream to which the current tuple will be emitted. Another overload takes one more seqId parameter, which is used to identify the tuple for ack or fail.

```
// Emit values to the default stream
public abstract void Emit(List<object> values);

// Emit values to specified stream
public abstract void Emit(string streamId, List<object> values);

// For non-transactional Spout only which support Ack. Emit values to
specified
// stream, and ack is required to this tuple, by using the unique Sequence
Id.
public abstract void Emit(string streamId, List<object> values, long seqId);

// For non-transactional Bolt only which support Ack. Emit values to
specified stream,
// and ack is required to this tuple, by using the input tuples as the
anchors.
public abstract void Emit(string streamId, IEnumerable<SCPTuple> anchors,
List<object> values);
```

Topology Builder

To build a topology in C#, you have to use the TopologyBuilder class. And it should be provided to Storm using another class that inherits class TopologyDescriptor. TopologyDescriptor has abstract method, GetTopologyBuilder. You should return your TopologyBuilder object from this method after configuring your topology correctly. This is how the plumbing code works in SCP.NET on a Storm cluster.

There are two main methods available on TopologyBuilder class: SetSpout and SetBolt.

Let's first examine usage of the SetSpout method. Here, let's assume there is an object of the TopologyBuilder class named topologyBuilder.

```
topologyBuilder.SetSpout(
    spoutName: "myspout",
createDelegate: MySpout.Get,
    outputSchema: new Dictionary<string, List<string>>() {
        {Constants.DEFAULT_STREAM_ID, new List<string>(){"field1",
"field2"}}
    },
parallelismHint: 1,
    customConfigFile: "Custom.config",
    enableAck: true);
```

Let's examine each line of code.

- The first line calls the SetSpout method.

- Then you pass a parameter. The first parameter is the spout name, which will appear in the Storm UI and logs.

- The second parameter is a delegate to the Get method, which should be defined in the MySpout class and return a MySpout object.

- The third parameter is the output schema (outputSchema). Here, you can specify the emitted tuple's stream and field names. You can specify multiple streams and their respective output schemas.

- The fourth parameter is parallelism hint (parallelismHint) to the Storm, which defines how many instances of the spout can be created.

- The fifth parameter is the custom configuration file (customConfigFile), where you can specify other parameters that this component should consider.

- In the final sixth parameter, select whether you want to enable acknowledgement (enableAck) for the emitted tuple or not.

Next is the SetBolt method. Again, let's assume there is an object of the TopologyBuilder class named topologyBuilder.

```
topologyBuilder.SetBolt(
    boltName: "mybolt",
    createDelegate: Splitter.Get,
    outputSchema: new Dictionary<string, List<string>>() {
        {Constants.DEFAULT_STREAM_ID, new List<string>(){"field1", "field2"}}
    },
    parallelismHint: 1,
    customConfigFile: "Custom.config",
    enableAck: true)
        .shuffleGrouping("myspout);
```

The SetBolt method is almost similar to the SetSpout method in terms of parameters, so I won't discuss it again. One thing you should notice is the shuffleGrouping method call. If you remember from the field/stream grouping section, you have different grouping on bolts. One of them is shuffle grouping. You can specify other groupings as well, like allGrouping, fieldsGrouping, and globalGrouping, and the tuples are diverted accordingly.

Using the Acker in Storm

There are three ways a topology can handle a message: non-transactional with no ack, non-transactional with ack, and transactional. (Ack stands for acknowledgment.) Each topology implementation has its own pros and cons. Let's discuss them all to understand when to use one instead of the others.

Non-Transactional Component Without Ack

Let's assume there is a topology with a spout and two bolts. Stream passes through components in a series. You send a stream through a topology, and if any of the bolts fail to process any message/tuple, then you don't have a mechanism to know about this failure. In the event of a failure in this kind of topology, you have potential data loss. This is at-most-once processing semantic, which means that if you try to process every tuple once, some of them are processed and some of them aren't. This is fine if you are reading temperature-sensor data 10 times every second, so even if you miss one or two readings in between, you are fine. So, use this kind of topology when data loss is not a problem and you can function properly if some loss occurs.

Non-Transactional Component with Ack

As discussed in the last section, if you are not comfortable with data loss, and you want systems that are more resilient, then there are a couple of options to guarantee message processing throughout the topology. To help you guarantee at-least-once processing semantics, there is a component called acker in Storm. When a bolt is done processing, it issues ack to acker, which informs the spout. A spout keeps a cache of information about the tuples requiring ack. Once the spout receives an ack for a tuple, it removes the tuple from the cache. Now assume that out of two bolts, one sends ack but the other fails. This informs the spout to replay the message. In this case, a Bolt processes the tuple more than once; hence, you need to have your own logic in a bolt to avoid this. But topology ensures that each tuple is processed at least once.

Transaction Component

Transactional component topology provides exactly-once semantics, as it guarantees that the message is processed once and only once. To build such a topology, you need to use the `TransactionalTopologyBuilder` class. These topologies are similar to non-transactional with ack topology; the difference is that the end bolt in topology is a committer bolt. It commits to the transaction and makes sure that the message is stored or processed successfully.

Building Storm Application in C#

When it comes to building a Storm application in C#, you need to ensure that you have all the prerequisites, listed as follows:

- Visual Studio 2012 with Update 4, 2013 with Update 4 (Community or higher), or 2015 (Community or higher)

- Azure SDK 2.9.5 or later

- HDInsight tool for Visual Studio (`http://bit.ly/2lhvW4U`)

Let's define the application that you are trying to build. You want to find influential tweets from the past hour or the past day. For this, you need to access the Twitter stream; hence, you will build a TwitterSpout. But before that, you need to know which tweets are the influential ones. For that, you take a very simple and naïve approach: you just count the sum of the number of retweets and the number of likes for a tweet. The tweets with the highest score qualify as influential tweets. To calculate the top tweet, you create two bolts: the first one calculates the total and finds the top 10 tweets during a window of time, let's say 5 seconds, and then forwards it to the second bolt, which saves it into a SQL database or a Hive table. And from there, you can fetch data directly into a dashboard or PowerBI for visualization.

So, that is the idea for the final application. Let's get started with the coding. After installing all the prerequisites, open Visual Studio. Go to File ä New ä Project. Select Azure Data Lake ➤ Storm (HDInsight). You see the different Storm starter templates, as shown in Figure 7-10. Note that this is the view from Visual Studio 2015.

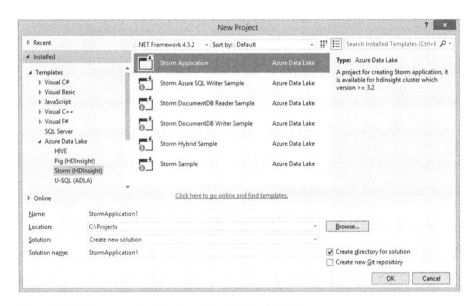

Figure 7-10. *Storm application templates in Visual Studio*

Use the Storm Application template, which provides a basic spout, bolt, and a topology class. Also, this template adds the required Nuget packages to the project, including SCP.NET. The Storm cluster also has SCP installed on it. At the time of writing, the default template added SCP.NET version 0.10.0.6, but there is an updated version 1.0.1 available for the package. Depending on the Storm version you have on your cluster, you need to install a SCP.NET package. Table 7-1 describes the versions of SCP.NET that work with particular versions of Storm and HDI.

Table 7-1. HDI, Storm, and SCP.NET Compatibility Matrix

HDInsight version	Apache Storm version	SCP.NET version
3.3	0.10.x	0.10.x.x
3.4	0.10.x	0.10.x.x
3.5	1.0.x	1.0.x.x

Before you start TwitterSpout, you need to have the Twitter app to connect to the Twitter stream. The following procedure explains how to create the new Twitter app. If you already have it, then you can skip this section.

1. Navigate to `https://apps.twitter.com` and log in with your Twitter account.

2. On the Application Management page, click the Create New App button. This takes you to the Create New Application page.

3. Fill out the new application information, including the unique app name, description, website, and optional callback URL.

4. Read and accept the developer agreement and hit the Create Your Twitter Application button. If all goes well, you will be redirected to your application page.

5. Navigate to the Keys and Access Tokens tab and copy Consumer Key (API Key) and Consumer Secret (API Secret).

6. For a new application, you have to generate access tokens. On the same page, there is an Access Token section. Hit the Create Access Token button and the page reloads with Access Token and Access Token Secret. Copy them and keep aside for TwitterSpout.

By using consumer key/secret and access token/secret, you can make an HTTP request to get tweets, but it would be a long process. There are ready-made libraries to access Twitter API easily. You will use the open source Tweetinvi library (`https://github.com/linvi/tweetinvi`).

Let's now create TwitterSpout.

1. Rename the existing spout class created by the Visual Studio template with TwitterSpout. This gives you `ISCPSpout` implementation.

2. After creating a spout class, tap into the Twitter stream. For that, add TweetinviAPI Nuget in the project. And then add a using statement for the `Tweetinvi` and `Tweetinvi.Models` in the class.

3. Next, add a StartStream method, which uses Twitter application tokens and assigns credentials to Tweetinvi for later use.

4. Create a filtered stream and add a filter on the language, because you are only interested in the English-language tweets.

5. Add a TweetReceived event handler and start the stream by calling the StartStream method. Whenever you receive a tweet, you add it to the queue. When the NextTuple method is called, you can pop a tweet from the queue and emit it to the downward stream.

The following is the code to configure Tweet API to receive tweets.

```
Queue<ITweet> queue = new Queue<ITweet>();
private void StartStream()
{
    Auth.SetUserCredentials(
        ConfigurationManager.AppSettings["ConsumerKey"],
        ConfigurationManager.AppSettings["ConsumerSecret"],
        ConfigurationManager.AppSettings["AccessToken"],
        ConfigurationManager.AppSettings["AccessTokenSecret"]);

    var stream = Tweetinvi.Stream.CreateSampleStream();
    stream.AddTweetLanguageFilter(LanguageFilter.English);
    stream.TweetReceived += (s, e) =>
    {
        if (e.Tweet.IsRetweet)
            queue.Enqueue(e.Tweet.RetweetedTweet);
    };
    stream.StartStream();
}
```

The following is the rest of the TwitterSpout code.

```
public class TwitterSpout : ISCPSpout
{
    private Context context;
    Thread listenerThread;

    long seqId = 0;
    Dictionary<long, ITweet> cache = new Dictionary<long, ITweet>(10000);
    private bool enableAck = false;

    public static List<Type> OutputSchema = new List<Type>() {
    typeof(SerializableTweet) };
```

```csharp
public static List<string> OutputSchemaName = new List<string>() {
"SerializableTweet" };

public TwitterSpout(Context ctx)
{
    this.context = ctx;

    Dictionary<string, List<Type>> outSchema = new Dictionary<string,
    List<Type>>();
    outSchema.Add("default", OutputSchema);
    this.context.DeclareComponentSchema(new ComponentStreamSchema
    (null, outSchema));

    // Get pluginConf info and enable ACK in Non-Tx topology
    if (Context.Config.pluginConf.ContainsKey(Constants.
    NONTRANSACTIONAL_ENABLE_ACK))
    {
        enableAck = (bool)(Context.Config.pluginConf
                [Constants.NONTRANSACTIONAL_ENABLE_ACK]);
    }
    Context.Logger.Info("enableAck: {0}", enableAck);

    listenerThread = new Thread(new ThreadStart(StartStream));
    listenerThread.Start();
}

public static TwitterSpout Get(Context ctx, Dictionary<string, Object>
parms)
{
    return new TwitterSpout(ctx);
}

public void NextTuple(Dictionary<string, Object> parms)
{
    if (queue.Count > 0)
    {
        var tweet = queue.Dequeue();
        cache.Add(seqId++, tweet);

        this.context.Emit(Constants.DEFAULT_STREAM_ID,
            new Values(new SerializableTweet(tweet)), seqId);
        Context.Logger.Info("Emit: {0}, seqId: { 1}", tweet.FullText,
        seqId);
    }
}
}
```

The TwitterSpout class defines SerializableTweet as the output schema. The SerializableTweet class is as follows. It needs to be serializable to make sure that it can be passed between components, and if required, it can work in hybrid topology as well.

```
[Serializable]
public class SerializableTweet
{
    public string Text { get; set; }
    public long Id { get; set; }
    public int RetweetCount { get; set; }
    public int FavoriteCount { get; set; }
    public decimal Score
    {
        get
        {
            return (RetweetCount + FavoriteCount);
        }
    }

    public SerializableTweet()
    {
        // For searialization and deserialization
    }

    public SerializableTweet(ITweet tweet)
    {
        this.Text = tweet.FullText;
        this.Id = tweet.Id;
        this.RetweetCount = tweet.RetweetCount;
        this.FavoriteCount = tweet.FavoriteCount;
    }

    public override string ToString()
    {
        return $"{Id.ToString()}:{Text}:Retweet-{RetweetCount}:Likes-
        {FavoriteCount}:Score-{Score}";
    }
}
```

Next, you need to create a bolt to find the top 10 tweets. Also, this bolt needs to find the top tweets during a 5-second period, and then emit those tweets to Azure SQL bolt to store them. You are going to use another feature of Storm, which is the Tick stream. The first bolt will subscribe to multiple streams: one coming from Spout and another one emitting a tuple every tick interval. Tick stream is special; it sends a tuple after the defined time interval. The following is the code for our first bolt, TopNTweetBolt. You can find the complete code on GitHub at http://bit.ly/2np3GeV.

```
public class TopNTweetBolt : ISCPBolt
{
    bool enableAck = false;
    private Context context;
    List<SerializableTweet> tweetCache = new List<SerializableTweet>();

    public static List<Type> OutputSchema = new List<Type>() {
    typeof(SerializableTweet) };
    public static List<string> OutputSchemaName = new List<string>() {
    "SerializableTweet" };

    public TopNTweetBolt(Context ctx, Dictionary<string, Object> parms)
    {
        this.context = ctx;

        // Input Schemas
        Dictionary<string, List<Type>> inSchema = new Dictionary<string,
        List<Type>>();
        // Default stream
        inSchema.Add(Constants.DEFAULT_STREAM_ID, TwitterSpout.OutputSchema);
        // Listen to the Tick tuple stream
        inSchema.Add(Constants.SYSTEM_TICK_STREAM_ID, new List<Type> {
        typeof(long) });

        // Output Schema to new stream named TopNTweets
        Dictionary<string, List<Type>> outSchema = new Dictionary<string,
        List<Type>>();
        outSchema.Add("TOPNTWEETS_STREAM", OutputSchema);

        this.context.DeclareComponentSchema(new
        ComponentStreamSchema(inSchema, outSchema));

        //If this task excepts acks you need to set enableAck as true in
        TopologyBuilder for it
        if (Context.Config.pluginConf.ContainsKey(Constants.
        NONTRANSACTIONAL_ENABLE_ACK))
        {
            enableAck = (bool)(Context.Config.pluginConf
                [Constants.NONTRANSACTIONAL_ENABLE_ACK]);
        }
        enableAck = true;
    }

    public static TopNTweetBolt Get(Context ctx, Dictionary<string, Object>
    parms)
    {
        return new TopNTweetBolt(ctx, parms);
    }
```

```
int totalAck = 0;
public void Execute(SCPTuple tuple)
{
    var isTickTuple = tuple.GetSourceStreamId().Equals(Constants.SYSTEM_
    TICK_STREAM_ID);
    if (isTickTuple)
    {
        // Get top 10 higest score tweets from last time window
        Context.Logger.Debug($"Total tweets in window: {tweetCache.
        Count}");
        var topNTweets = tweetCache.OrderByDescending(o => o.Score)
            .Take(Math.Min(10, tweetCache.Count)).ToList();

        // Emit it to TopNTweet Stream
        foreach (var tweet in topNTweets)
        {
            this.context.Emit("TOPNTWEETS_STREAM", new Values(tweet));
        }

        // Remove all existing data and wait for new one
        tweetCache.Clear();
    }
    else
    {
        try
        {
            // Process tuple and then acknowledge it
            SerializableTweet tweet = tuple.GetValue(0) as
            SerializableTweet;
            if (!tweetCache.Any(o => o.Id.Equals(tweet.Id)))
                tweetCache.Add(tweet);

            Context.Logger.Info(tweet.ToString());

            if (enableAck)
            {
                this.context.Ack(tuple);
                Context.Logger.Info("Total Ack: " + ++totalAck);
            }
        }
        catch (Exception ex)
        {
            Context.Logger.Error("An error occured while executing Tuple
            Id: {0}. Exception Details:\r\n{1}",
                tuple.GetTupleId(), ex.ToString());

            //Fail the tuple if enableAck is set to true in TopologyBuilder
            //so that the tuple is replayed.
```

```
            if (enableAck)
            {
                this.context.Fail(tuple);
            }
        }
    }
}
```

In the execute method, you first differentiate between the streams you received. If it is a tick tuple stream, then you emit the top 10 tweets to a new stream named "TOPNTWEETS_STREAM". The next bolt (AzureSqlBolt) in the topology subscribes to this new stream and saves all the tweets to an Azure SQL database (to create a new Azure SQL database, follow the guide at https://docs.microsoft.com/en-us/azure/sql-database/sql-database-get-started). From there, perhaps the top tweets from the past hour or past day can be read on a browser or PowerBI.

The next code is from AzureSqlBolt. You will use Visual Studio's AzureSqlBolt template of to generate a new bolt AzureSqlBolt. The following only lists the execute method. The rest of the code is fairly straightforward.

```
public class AzureSqlBolt : ISCPBolt
{
...

    public void Execute(SCPTuple tuple)
    {
        try
        {
            SerializableTweet tweet = tuple as SerializableTweet;
            Context.Logger.Info("SQL AZURE: " + tweet.ToString());

            List<object> rowValue = new List<object>();
            rowValue.Add(tweet.Id);
            rowValue.Add(tweet.Text);
            rowValue.Add(tweet.RetweetCount);
            rowValue.Add(tweet.FavoriteCount);
            rowValue.Add(tweet.Score);
            rowValue.Add(DateTime.UtcNow);
            Upsert(new List<int> { 1 }, rowValue);
        }
        catch (Exception ex)
        {
            Context.Logger.Error(
                "An error occured while executing Tuple Id: {0}. Exception
                Details:\r\n{1}",
                tuple.GetTupleId(), ex.ToString());
        }
    }

...
}
```

The final thing to do to make the code work is define the topology. The Program.cs file already has dummy code for this. Let's replace it with the following.

```
[Active(true)]
class Program : TopologyDescriptor
{
    public ITopologyBuilder GetTopologyBuilder()
    {
        TopologyBuilder topologyBuilder = new TopologyBuilder(
            "TwitterStreaming" + DateTime.Now.ToString("yyyyMMddHHmmss"));

        topologyBuilder.SetSpout(
            "TwitterSpout",
            TwitterSpout.Get,
            new Dictionary<string, List<string>>()
            {
                {Constants.DEFAULT_STREAM_ID, TwitterSpout.OutputSchemaName}
            },
            1, true);

        var boltConfig = new StormConfig();
        boltConfig.Set("topology.tick.tuple.freq.secs", "5");
        topologyBuilder.SetBolt(
            "TopNTweetBolt",
            TopNTweetBolt.Get,
            new Dictionary<string, List<string>>(){
                {"TOPNTWEETS_STREAM", TopNTweetBolt.OutputSchemaName}
            }, 1, true)
            .shuffleGrouping("TwitterSpout")
            .addConfigurations(boltConfig);

        topologyBuilder.SetBolt(
            "AzureSqlBolt",
            AzureSqlBolt.Get,
            new Dictionary<string, List<string>>(),
            1).shuffleGrouping("TopNTweetBolt", "TOPNTWEETS_STREAM");

        return topologyBuilder;
    }
}
```

One thing you might notice is that the first bolt has a bolt configuration and you set topology.tick.tuple.freq.secs parameter to 5. This means the tick stream will generate a tuple every 5 seconds and send it to this bolt.

Lastly, you can submit this topology to a Storm cluster. I'm using a Windows-based cluster, but Linux clusters are also supported. To submit a topology, right-click the project node, and then select Submit to Storm on HDInsight. It should open a dialog similar to what's shown in Figure 7-11.

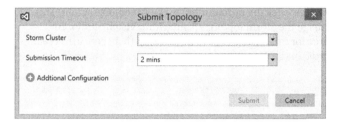

Figure 7-11. *Submit topology from Visual Studio*

You may be asked for credentials. If you have any Storm clusters in your subscription, the name of the cluster will appear in first in the drop-down. Select one of the Storm clusters and then click the Submit button. This generates the package file and submits it to the Storm cluster. Once it starts running, you can see the topology from Visual Studio only. Figure 7-12 shows what the topology looks like after running for few minutes.

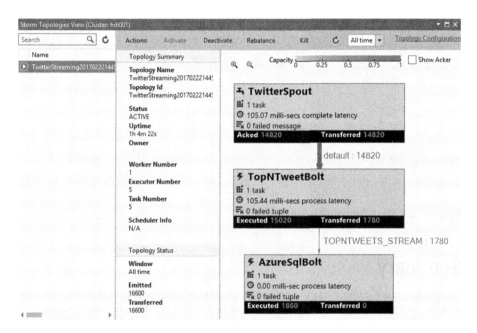

Figure 7-12. *Storm topology stats in Visual Studio*

When you click any of the topology components, you get a detailed view of that component. So, let's look at `TopNTweetBolt`. Figure 7-13 shows the bolt in a topology that has been running for a few minutes. It shows various stats about the component, input and output stats, and any errors that have occurred in it. You can change the time window by clicking the time shown in bolt stats, and the rest of the stats change accordingly.

And if you want to look at the logs, click the port number in the Executors section. This way, you can completely manage the Storm topology from within Visual Studio. To download the complete source for this example, go to `https://github.com/vinityad/InfluentialTweetWithStormHDInsight`.

Figure 7-13. *TopNTweetBolt summary*

Summary

Apache Storm is a reliable way to process unbounded streams of data. It is easy to program and work with. It provides low-latency processing pipelines for streams. It is fault tolerant and resilient. It is based on Hadoop, hence it leverages the scaling behavior of Hadoop. Storm topology can be built in any shape or size with spouts and bolts. Stream grouping provides sophistication on top of the already great architecture. And finally, HDInsight provides C# developers a way to program any real-time processing pipeline in Storm, leveraging existing knowledge of C#. Visual Studio integration makes it very easy for .NET developers to jump in and start developing, deploying, testing, and monitoring Storm app. So, far you have learned everything from batch processing to real-time analytics. But each thing requires the learning a new tool or technology.

The next chapter looks at Apache Spark, which covers most of the big data scenarios.

CHAPTER 8

■ ■ ■

Exploring Data with Spark

Apache Spark changed the landscape of big data and analytics when it came out. Developers welcomed it like nothing else. It quickly became the superstar from ascendant technology. It is one of the most active and contributing open source projects in the big data ecosystem. At the time of writing, there are more than 1000 contributors to the project. Many big data companies have started moving from MapReduce to Spark. And there is no single reason for them to do so. Spark provides improvements in handling data, and it is very easy to work with. Before Spark, if you wanted to do batch processing, interactive query, machine learning, and stream analytics, then you would have used multiple tools like MapReduce, Hive, Storm, and so forth. And maintaining such a system with a wide range of technologies is not easy. Apache Spark can handle all of these scenarios and makes developers' lives easy—one of the many reasons that Spark is so popular among the big data community.

This chapter first discusses Spark and the problems that it solves, and then looks at HDInsight's Spark offering. In this chapter, you also perform practical with Spark shell and Notebooks. You also learn concepts such as RDD (Resilient Distributed Datasets), DataFrames and Datasets, Spark SQL, Spark Streaming, and finally, you build a standalone Spark application in C#. Let's start with Spark overview.

Overview

Apache Spark is an open source, fast, in-memory data processing engine. It provides powerful and expressive development APIs, which allow the big data community to efficiently create and execute SQL workloads, machine learning tasks, and graph processing and streaming jobs. Also, it specifically boosts performance for iterative data access scenarios through in-memory computation capabilities. Spark running on Apache Hadoop YARN can leverage Spark's power to handle their big data workloads on distributed data in Hadoop. The YARN-based architecture provides a foundation that enables Spark to utilize resources available to process data while maintaining a consistent level of service and response.

Spark was originally developed by UC Berkeley's AMPLab in 2009. In 2010, it was open sourced as an Apache project. There are several advantages that Spark provides over MapReduce; the following are a few of them.

- Spark can work with a variety of workloads: batch processing, and real-time or graph data processing.

© Vinit Yadav 2017

V. Yadav, *Processing Big Data with Azure HDInsight*, DOI 10.1007/978-1-4842-2869-2_8

- Spark's in-memory computation gives 100 times faster performance and 10 times faster when running on disk compared to the MapReduce engine.

- Spark's comprehensive developer APIs are easy to use and reduce the complexity of writing code to run in a distributed fashion.

- Easily develops applications in your choice of programming language: Java, Scala, Python, or R (or C# and F# using Mobius).

- Offers out-of-the-box capabilities to work with machine learning problems.

- Works with a variety of data sources, such as HDFS, HBase, Casandra, Azure Blob storage, and so forth.

The Spark ecosystem contains Spark Streaming, Spark SQL, MLlib, and GraphX. The following are brief descriptions.

- **Spark Streaming**: Allows the handling and processing of streaming data. Essentially, it is microbatching a continuous stream over a time window. Continuous RDDs (explained later in this chapter) called DStream process real-time data. For more information about Spark Streaming, refer to `http://spark.apache.org/docs/latest/streaming-programming-guide.html`.

- **Spark SQL**: Spark provides running SQL queries on Spark datasets using traditional BI and visualization tools. Spark SQL can load data from various formats (text, JSON, Parquet, or from a database), transform it to RDD, and allow for ad hoc format-independent querying. Spark 2.0 supports SQL 2003.

- **Spark MLlib**: Spark's scalable machine learning library. MLlib contains many algorithms and utilities that cover almost all the basics of machine learning algorithms. MLlib uses Spark's in-memory processing capability to train and run models faster. For more information about Spark machine learning, go to `http://spark.apache.org/docs/latest/ml-guide.html`.

- **Spark GraphX**: Allows you to process graph data using Spark's RDD. It introduced a new graph abstraction with directed multigraph. In the graphs, properties are attached to each vertex and edge. Also, it supports a wide range of algorithms and builders to make graph analytics tasks easier.

Spark Architecture

Apache Spark uses a master/slave/worker architecture. A driver program runs on the master node and talks to an executor on worker node. Both the driver and the executors spawn their own JVM processes. Spark applications run as independent sets of processes, which is coordinated by the SparkContext object and created by the driver program.

Spark can run in standalone mode or on a cluster of many nodes. Hence, a cluster manager can be a standalone cluster manager, Mesos, or YARN. SparkContext talks to the cluster manager to allocate resources across applications. Spark acquires executors on nodes in the cluster, and then sends application code to them. Executors are processes that actually run the application code and store data for these applications. And finally, the driver sends tasks to executors, as shown in the flowchart in Figure 8-1.

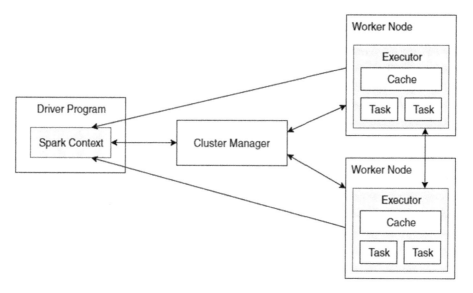

Figure 8-1. *Spark architecture*

The following are a few observations about the Spark architecture.

- A worker node spawns an executor for each application. It stays there as long as the application exists. Each executor runs a task in which there can be multiple threads. This isolates each application from each other. And any cached data is stored in memory or on disk, based on the options provided.

- Spark does not care about which cluster manager it is working with as long as it can acquire executors and communicate with them. Given that, running Spark becomes easy on the cluster manager, which also supports other applications (e.g., Mesos/ YARN).

- Driver programs need to communicate with worker nodes to schedule tasks on them, due to which it is advisable to have the worker and driver program in the same local area network.

The following are the cluster manager types available for Spark.

- **Standalone**: This cluster manager makes it easy to set up a cluster. It is already included in Spark to simplify the process further.

- **Apache Mesos**: A generic cluster manager type. It can also run the service application and MapReduce.

- **Hadoop YARN**: The de facto Hadoop 2 resource manager.

Spark clusters on HDInsight come with Anaconda libraries pre-installed. Anaconda provides close to 200 libraries for machine learning, data analysis, visualization, and so forth. Figure 8-2 shows the overall Spark architecture in HDInsight and the ecosystem structure discussed so far.

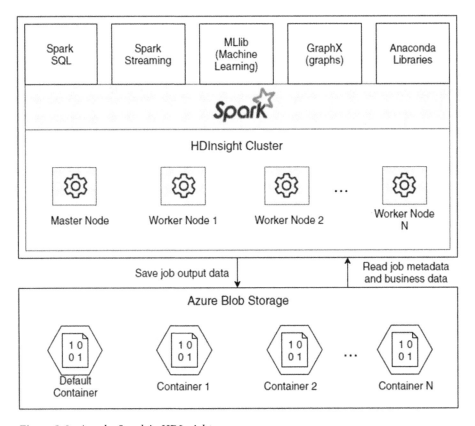

Figure 8-2. *Apache Spark in HDInsight*

Creating a Spark Cluster

Azure HDInsight provides many different cluster types, which were discussed in Chapter 2. To work with Spark, you need a head node and worker node. HDInsight provides a simple-to-deploy Spark cluster type that automatically creates all the nodes. There are multiple ways to create a Spark cluster (i.e., Azure portal, PowerShell and Azure Resource Manager templates). Let's look at an Azure Resource Manager template to create a Spark cluster.

The Azure Resource Manager (ARM) template is a JSON file with different resource and configuration options to deploy/provision/configure a resource in Azure. You will use the template available at `https://raw.githubusercontent.com/Azure/azure-quickstart-templates/master/ 101-hdinsight-spark-linux/azuredeploy.json`. The following procedure creates a Linux-based Spark cluster.

Open the following URI, which takes the ARM template URL and opens an easy-to-configure options page to create a Spark cluster.

```
https://portal.azure.com/#create/Microsoft.Template/uri/
https%3A%2F%2Fraw.githubusercontent.com%2FAzure%2Fazure-quickstart-
templates%2Fmaster%2F101-hdinsight-spark-linux%2Fazuredeploy.json
```

- **Subscription**: Select the Azure subscription to be used by your cluster.

- **Resource group**: You can choose to create a new resource group or to select an existing group.

- **Location**: Specify where your cluster and data will reside.

- **ClusterName**: A unique name for the cluster using `https://{clustername}.azurehdinsight.net`. In the template, you may not be able to select the Spark cluster version but HDInsight provides support to Spark 1.5.2 (HDI 3.3), 1.6.2 (HDI 3.4), 1.6.3 (HDI 3.5), 2.0.2 (HDI 3.5), and 2.1.0 (preview) (HDI 3.6) at the time of writing.

- **Cluster credentials**: The cluster's username and password to open the dashboard. These credentials are used to submit jobs and get into the cluster. The default username is admin.

- **SSH credentials**: The username and password to get onto a Linux machine.

Agree to the terms and conditions and click the Purchase button.

By default, this template provisions two worker nodes and two head nodes. Once you hit the Purchase button, it takes a few minutes to provision the cluster. You can browse the Ambari dashboard by going to `https://{ClusterName}.azurehdinsight.net` once provisioning is completed.

Spark Shell

Spark shell is an interactive shell to run ad hoc queries in Spark. This helps in learning Spark commands and requires almost no setup once the HDInsight Spark cluster is up. Spark shell is like a playground, where you can try different things, run queries, and get familiar with Spark's features.

Once you have provisioned your Spark cluster, you can get onto the Spark shell by using Putty, as discussed in earlier chapters. After you have successfully authenticated yourself, type **pyspark** in console, which creates a Spark shell and shows a screen like the one in Figure 8-3.

Figure 8-3. *Spark shell*

After you get into Spark shell, you can load data and start processing it. Let's assume that you want to count the number of non-empty lines in a text file. When you load Spark shell, it automatically creates a Spark context with the name sc. You can use SparkContext to load data, as shown in the following code snippet.

```
lines = sc.textFile('/example/data/gutenberg/davinci.txt')
```

Don't worry about uploading the text file; it should already be in your Azure Blob storage linked to the HDInsight cluster. The important thing to note in this snippet is that it uses the new line as a delimiter and creates an RDD (more on this later) with each row.

Now that you have your data, you want to remove all the empty lines, and perform a count afterward. (Don't worry about the syntax; you won't be using it for a real-world application).

```
filtered_lines = lines.filter(lambda x: len(x) > 0)
filtered_lines.count()
```

Executing all of these statements in Spark shell gives an output similar to what's shown in Figure 8-4.

```
  \ \/ _ \/ _ '/ _/ ' /
 /_ / .__/\_,_/_/ /_/\_\   version 2.0.2.2.5.4.0-121
    /_/

Using Python version 2.7.12 (default, Jul  2 2016 17:42:40)
SparkSession available as 'spark'.
>>> lines = sc.textFile('/example/data/gutenberg/davinci.txt')
>>> filtered_lines = lines.filter(lambda x: len(x) > 0)
>>> filtered_lines.count()
[Stage 0:>
[Stage 0:==============================>

25163
>>>
```

Figure 8-4. *Spark job output*

Spark RDD

Spark's core concept is Resilient Distributed Dataset (RDD); everything else revolves around it. RDD is a fault-tolerant collection of elements. RDD can be operated on in parallel. As the name suggests, it is a resilient and distributed collection of records, which can be at one partition or more, depending on the configuration. RDD is an immutable distributed collection of objects, which implies that you cannot change data in RDD but you can apply transformation on one RDD to get another one as a result. It abstracts away the complexity of working in parallel. You can use a higher-level programming interface (API) to do processing without much focus on parallelism, which is handled by Spark itself.

You can create RDD either by parallelizing an existing collection or loading an external dataset, such as a shared file system, HDFS, HBase, or any data source offering a Hadoop input format.

Let's decompose the name,

- **Resilient**: Fault tolerant, comes from an RDD linage graph, able to re-compute when it has missing records or damaged partitions due to node failures.

- **Distributed**: Data resides on multiple nodes in a cluster.

- **Dataset**: A collection of partitioned data with a key-value pair or primitive values called tuples. Represents the records of data you work with.

Other than the traits embedded in the name, the following are additional traits of RDD.

- **Immutable**: RDD never changes once created; they are read-only and can only be transformed to another RDD using transformation operations.

- **Lazy evaluated**: In the example shown in the Spark shell section, you first load data into RDD, then apply a filter on it, and ask for a count. If you noticed, there were no computations shown in shell after the filter transformation and the actual job only kicks off after you perform an action on RDD. That implies that RDD is not transformed until an action is executed, which actually triggers the execution.

- **In-memory**: Spark keeps RDD in memory as much size as it can and for as long as possible.

- **Typed**: RDD records are strongly typed, like Int in RDD[Int] or tuple (Int, String) in RDD[(Int, String)].

- **Cacheable**: RDD can store data in a persistent storage-like memory, which is the default and preferred for better performance, or on disk.

- **Partitioned**: Data is split into a number of logical partitions based on multiple parameters, and then distributed across nodes in a cluster.

- **Parallel**: RDDs are normally distributed on multiple nodes, which is the abstraction it provides; after partitioning, it is acted upon in parallel fashion.

- **Location aware**: RDDs has location preference, Spark tries to create them as close to data as possible provided resources are available.

Determining partitions for an RDD is a distributed process;, it tries to achieve an even data distribution and to keep data as close to the source as possible (data locality). Partitions are created in a fixed number based on logical chunks of data. This division is for processing only; internal storage is not divided. Partitions are the units of parallelism in Spark. Spark automatically divides data into a number of partitions defined as per configuration and other parameters, but if you think Spark is dividing it incorrectly by over or under partitioning, then you can provide your own number of partitions in transformations. Spark always tries to avoid sending data to another node—a process called *RDD shuffling*. It usually follows one-to-one mapping between physical data chunks to partitions (e.g., HDFS partitions).

There are two kinds of operations you can perform on RDDs: transformations and actions. Transformations are lazy operations and return another RDD. Spark accumulates all such operations performed on RDDs and runs them as a job when an action is called on the final RDD. Actions, on the other hand, trigger the computation and return a value.

Let's examine both operations in more detail.

RDD Transformations

Transformations take an RDD and generate one or more RDDs (e.g., `filter`, `map`, `flatmap`, `join`, `reduceByKey`). Transformations cannot change an RDD because it is immutable; hence, they generate new RDDs from the input RDD. When you apply transformations to an RDD, you are essentially building up RDD lineage graph. Spark does this for each transformation that you apply, which is how Spark can re-compute any intermediate RDD in the event of a failure. A lineage graph is the way that Spark achieves fault tolerance.

Transformations are lazy evaluated, and only executed when you perform actions on them. Lazy evaluation gives Spark a chance to build the optimum plan to execute them with a minimum of passes on data. Also, certain transformations can be pipelined, which is an optimization Spark does to improve the performance of operations.

There are two types of transformations: narrow and wide. In narrow transformations, each child RDD partition depends on—at most—one parent RDD partition. That means the task can be executed locally without shuffling data. Another way to think of this is that any row in a child RDD depends on only one row of the parent RDD. It is also called a *narrow dependency*. Example transformations are `map`, `flatmap`, `filter`, and so forth. Spark groups these transformations by a method called *pipelining*.

Wide transformations, on the other hand, can have multiple child partitions, depending on one partition of the parent RDD. In other words, data required to compute the output records in a single partition may reside in many partitions of the parent RDD (e.g., `groupByKey` and `reduceByKey`). All the tuples with the same key must end up in the same partition and must be processed by the same task.

The following describes a few of the most common transformations.

- map(func): Generates a new RDD after passing each element of the source RDD through the func function. Figure 8-5 shows a sample map function that converts an input string into lowercase.

Figure 8-5. *Map transformation*

- flatmap(func): Similar to map, all elements of a parent RDD are passed through the func function, which may return multiple results. All of those results are then flattened to form a resultant RDD. Figure 8-6 shows a source RDD with statements. You want to have each word as a resultant RDD, applying a flatmap with a split function on a space gives the required output.

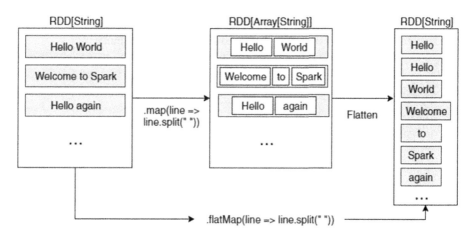

Figure 8-6. *Flatmap transformation*

- filter(func): Returns a new dataset formed by passing the source element through the func function and selecting those elements where the func function returns true.

- `mapPartitions(func)`: Similar to map, but runs on each partition; hence the `func` function should be of type `Iterator<T> => Iterator<U>` when running on an RDD of type T.

- `sample(withReplacement, fraction, seed)`: Generates a new RDD from the `fraction` fractions of the data, with or without replacement, and using `seed` as random number generator.

- `union(otherDataset)`: Creates a new dataset containing all the elements from the source and `otherDataset`.

- `intersection(otherDataset)`: Generates a new RDD that contains the elements from the datasets that are in common.

- `distinct()`: As the name suggests, returns the distinct element from the source RDD.

- `groupByKey()`: When called on a `RDD(tuple(k, v))`, it returns `RDD(tuple(k, iterable<v>))`. The default level of parallelism depends on the number of partitions of the parent RDD, but you can also pass an optional `numTasks` argument to set a different number of tasks.

- `reduceByKey(func)`: Similar to `groupByKey`, but optimized for aggregation. In this transformation, data is not shuffled at the beginning, because it knows that there is a reduce operation; hence, it applies a `reduce` on each partition and only sends the result of the `reduce` function over the network as a shuffle. This reduces significant network traffic and improves overall performance. Figure 8-7 shows a sample `reduceByKey` operation where from each partition data is reduced and then the shuffle operation is carried out to give a final result.

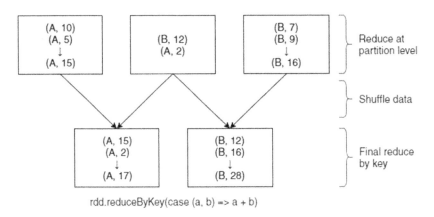

Figure 8-7. reduceByKey transformation

- `aggregateByKey(zeroValue)(seqOp, combOp)`: When called on `RDD(tuple(k, v))`, it returns `RDD(tuple(k, u))`, where the value of each key is aggregated using the given combined function and natural "zero" value.

- `join(otherDataset)`: Combines the two datasets based on a key. When called with `RDD(tuple(k, v))` and `RDD(tuple(k, w))`, it generates a resultant `RDD(tuple(k, (v, w)))`.

- `repartition(numPartitions)`: If you think that Spark is dividing data into the wrong number of partitions, so you want to change it to fewer or more partitions, use this transformation with a new partition value. Please note that this always shuffles data over the network.

- `coalesce(numPartitions)`: Decreases the number of partitions in the RDD to `numPartitions`. This is useful for running an operation more efficiently after filtering a large dataset.

RDD Actions

Actions are operations that return a value to the driver program. Actions produce non-RDD values. Once applied on an RDD, they trigger the actual execution of all the transformation applied before it. To put it simply, actions evaluate the RDD lineage graph. Note that if you are supposed to execute two or more actions on an RDD, then you should cache the RDD for better performance (more on caching in coming sections). The following are a few of the most commonly used actions.

- `reduce(func)`: Aggregates the elements of the dataset using a `func` function. The function takes two arguments and returns the one that should be of the same type as input arguments. The function should be commutative and associative, making sure that it can execute correctly in parallel.

- `collect()`: This action asks each executor to return a value to the driver program as an array. You use this action usually after a filter or other operations that return a significantly small subset of data.

- `count()`: Returns the number of elements in the dataset.

- `take(n)`: Returns the first n elements from the dataset.

- `first()`: Similar to take(1), returns the first element from the dataset.

- `takeSample(withReplacement, num, [seed])`: Returns random num elements from the dataset with an optional seed as the random number generator.

- `takeOrdered(n, [ordering])`: Returns the first n elements from the dataset using either their natural order or a custom comparator.

- `saveAsTextFile(path)`: Creates a text file with all the elements from the dataset in a given path. It can be local machine's file system or any other Hadoop-supported file system. Spark will call toString on each element to convert it to a line of text in a file.

- `countByKey()`: Only available on RDDs of type Tuple(k, v). Returns tuple(k, int) with the count of each key.

- `foreach(func)`: Executes the `func` function on each element of the dataset. This is usually done to print or update accumulator variables or to interact with external storage systems.

Shuffle Operations

Certain operations in Spark trigger an event known as the *shuffle* to redistribute data across a cluster so that it can be grouped differently across partitions. In a shuffle, data is copied between executors and nodes, which make it an expensive operation to perform and have performance impact.

To understand a shuffle and why it is required, assume a situation where you have server log data and you want to count that number of errors generated each hour throughout the day. So, you take all error records and get their timestamp, generate a new element with just hour of the day as the key, and have "1" as the value. Next, you combine all the same key values. But now assume you have data distributed all over your cluster in different machines. The same key is distributed on different nodes, making it impossible to count the total without getting all the same key data in one place. Hence, in this situation, the data needs to be transmitted to form different partitions, which is a shuffle operation.

Not every operation causes a shuffle. Operations the can cause a shuffle include repartition operations, such as `coalesce` and `repartition`; ByKey operations (not counting), such as `groupByKey` and `reduceByKey`; and join operations, such as `cogroup` and `join`.

A shuffle is an expensive operation because it involves moving data over a network, disk I/O, and data serialization. In a shuffle, Spark uses sets of tasks to organize data: map tasks to organize data and reduce tasks to aggregate data. Please note this map and reduce are not related to Spark's operations. Internally, Spark stores all the result of map operations in memory until they can't fit. Afterward, they are sorted based on target partition and written to a single file. The reduce task reads the sorted blocks.

A few shuffles can cause a lot of data to be stored in memory. And when data cannot fit into the memory, Spark spills the data to disk, which adds more performance hits due to I/O and increased garbage collection. Besides, Spark also generates a lot of intermediate files on the disk. These files are persisted until the related RDDs are in use and not garbage collected. This is done to avoid recalculation of the shuffle when RDD is re-computed from lineage. Storing files may pile up and take up a large amount of disk space if garbage collection doesn't happen frequently or if the application retains reference to RDDs. Either way, it wastes disk space temporarily. But if you feel a shuffle is not optimum, you can control this behavior by using a variety of configuration parameters.

Persisting RDD

Spark's main selling point is persisting a dataset in memory. You can ask Spark to store an RDD for a subsequent operation in memory so that you can reuse that data again in another action without the overhead of computing the RDD again. This greatly improves the performance in iterative algorithms and fast interactive use. Whenever you ask to cache/persist an RDD, each partition stores their part of the data in their memory and reuses it when needed.

To store an RDD, you can use the persist() or cache() methods on it. When an action operation is performed on an RDD for the first time, the RDD is stored in the memory of the nodes. If for any reason the RDD is corrupt or lost, it can be recomputed easily from a lineage graph, as Spark's cache is fault tolerant.

Spark also provides different storage levels, such as the persist dataset on disk, in memory as serialized Java objects or replicate across nodes. You can set the storage level in the persist() method. The cache() method is equivalent to persist with a MEMORY_ONLY storage level. The following are all the storage levels available in Spark.

- MEMORY_ONLY: This is a default level and in this level, RDD is stored as a deserialized Java object in the JVM. And if a dataset cannot fit the memory, then it is recomputed as and when needed.

- MEMORY_AND_DISK: Similar to MEMORY_ONLY, but the difference is that if the RDD doesn't fit in memory, then it stores partitions on a local disk.

- MEMORY_ONLY_SER: Data is stored in memory but as a serialized Java object. This greatly reduces the space of the data but it is more CPU-intensive to read.

- MEMORY_AND_DISK_SER: Similar to MEMORY_ONLY_SER, but it stores data on a local disk, which cannot fit into memory.

- DISK_ONLY: This option makes Spark store all RDDs on a local disk.

- MEMORY_ONLY2 and MEMORY_AND_DISK2: The same as their counterparts, but replicate partition on two cluster nodes.

- OFF_HEAP: This option is similar to MEMORY_ONLY_SER but it stores data in off-heap memory. To enable this storage level, off-heap memory needs to be enabled.

The in-memory option is best if your data fits in memory. This option is the default. It is also the most CPU-efficient and runs as fast as possible. If you can't fit it in memory, then use MEMORY_ONLY_SER to make objects smaller and fit in memory. The local disk option should only be used when compute options are very expensive or if they are generated after scanning large amounts of data. Two replications should only be used when you need fast fault recovery.

After all of this, if you know you no longer need the data in application, then you can remove the persisted RDD by using the unpersist() method; otherwise, RDD is removed in a least recently used (LRU) fashion.

Spark Applications in .NET

Spark doesn't provide out-of-the-box support for C# applications. But thanks to Microsoft, you have an almost complete API for Spark in C# as well. A project called Mobius (https://github.com/Microsoft/Mobius) is an open source C# language binding to Apache Spark. Mobius enables the implementation of the Spark driver program and data processing operations in the .NET framework in language like C# and F#. With Mobius, developers can implement C#-based Spark applications for batch, interactive, and stream processing, which makes C# a first-class citizen in Spark app development.

Let's explore Mobius in detail to understand how it works. Figure 8-8 shows .NET and Spark interaction. You can write the complete application in .NET and you don't have to write a single line of code in Java or Scala. You interact with the Mobius API, which depends on Scala/Java public APIs and the Python API.

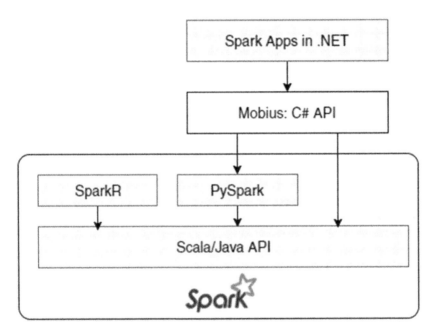

Figure 8-8. *.NET & Spark*

To develop a Spark application in .NET, you open Visual Studio, create a console app, add a Mobius-related Nuget, and use the sparkclr-submit command to submit your .exe to run inside the Spark installation. So, this is an obvious question to anyone familiar with .NET and Spark: Spark has all JVM process, driver and executors are both JVM processes, and .NET exe requires CLR to run, then how do both of these environments work on a Spark installation? The answer to the question is shown in Figure 8-9.

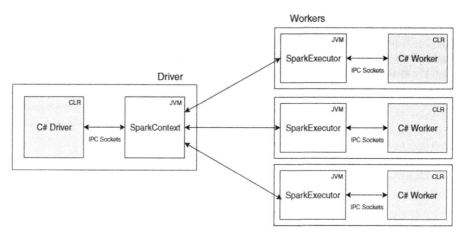

Figure 8-9. *Mobius and Spark*

Figure 8-9 shows a standard Spark setup, with the driver program communicating with executors. When you submit a .NET executable to this setup through Mobius and C#, the driver creates a SparkContext. It creates all the objects related to it as well, which JVM would have created. And if there is a C# user -defined function or any C#-specific functionality in the processing pipeline, then executor launches C# worker, because C# code can only be executed inside CLR. But please note that a C# worker is launched on demand, which means if there is any C-specific code, then only the executor will launch a C# worker' otherwise, it will run completely inside the JVM process. For example, if you are just reading a file and doing counts on it, then this uses no C# functionality; so it executes without a C# worker completely inside JVM.

You might think that this is CLR process–based, hence it only works in Windows. That is not the case. On Linux machines, Mobius uses the Mono framework to do the same work. That means you can use Mobius on any installation of Spark where you have Mono or you are running on Windows.

Let's build a simple word count program in C# and then run it in local mode and cluster mode. Let's build the code and then submit it to different environments.

Developing a Word Count Program

You will create a word count program in C# to read a text file and count word frequency in it. The following procedure explains how to create a word count program.

1. Open Visual Studio (2012 or higher) and create a C# console application. Name it SparkClrWordCount.

2. Install a Mobius package by executing the install-package Microsoft.SparkCLR command in a package manager console. At the time of writing, the latest version is 2.0.200, which works with Spark 2.0.2. After installing Nuget and updating to latest version, package.config looks like the following.

```xml
<?xml version="1.0" encoding="utf-8"?>
<packages>
<package id="log4net" version="2.0.7" targetFramework="net45" />
<package id="Microsoft.SparkCLR" version="2.0.200"
targetFramework="net45" />
<package id="Newtonsoft.Json" version="9.0.1"
targetFramework="net45" />
<package id="Razorvine.Pyrolite" version="4.18.0"
targetFramework="net45" />
<package id="Razorvine.Serpent" version="1.18.0"
targetFramework="net45" />
</packages>
```

3. Load the text file using the SparkContext.TextFile
    method. The path of the text file comes from a command-
    line argument. Once loaded as RDD, apply a split by space
    by using the Split method with a space character as the
    parameter, convert it to a key-value pair using the Map
    method, and then use the ReduceByKey method to calculate
    word count. The following is code from the Program.cs file's
    Main method.

```csharp
public static int Main(string[] args)
{
    LoggerServiceFactory.SetLoggerService(Log4NetLoggerService.Instance);
    ILoggerService Logger = LoggerServiceFactory.GetLogger(typeof(WordCount
    Example));

    if (args.Length < 1)
    {
        Console.Error.WriteLine("Usage: SparkClrWordCount  <file>");
        return 1;
    }

    var sparkContext = new SparkContext(new SparkConf().SetAppName("MobiusW
    ordCount"));

    try
    {
        RDD<string> lines = sparkContext.TextFile(args[0]);
        RDD<Tuple<string, int>> counts = lines
            .FlatMap(x => x.Split(' '))
            .Map(w => new Tuple<string, int>(w, 1))
            .ReduceByKey((x, y) => x + y);

        foreach (var wordcount in counts.Collect())
```

```
    {
        Console.WriteLine("{0}: {1}", wordcount.Item1, wordcount.Item2);
    }
}
catch (Exception ex)
{
    Logger.LogError("Error performing Word Count");
    Logger.LogException(ex);
}
sparkContext.Stop();
return 0;
}
```

In the Main method, you first create the SparkContext, with the AppName configuration option as MobiusWordCount. Using the same context object, load the text file as RDD. And then apply different transformations to the RDD (i.e., flatmap, map, and reduceByKey). Finally, in the for loop, ask for all the elements by applying a Collect action, which actually triggers the computation and returns the result to the driver program, where you just print it out on a console.

Next, you look at how to run this in local mode and on an HDInsight cluster.

Running in Local Mode

To run a C# program in local mode, the following lists the prerequisites you should have in your machine. Note that I'm assuming you have a Windows machine and want to run the word count program in the same box. To run it on a Linux box, use the instructions provided at GitHub (https://github.com/Microsoft/Mobius/blob/master/notes/linux-instructions.md). You should have at least 4GB of RAM for smooth functioning and Windows 8 or higher.

- **JDK**: You need to install JDK 7u85 or 8u60 (or higher) from OpenJDK (http://www.azul.com/downloads/zulu/zulu-windows/) or Oracle JDK (from http://www.oracle.com/technetwork/java/javase/downloads/index.html). Add a system-level environment variable, JAVA_HOME, which should point to the JDK folder (i.e., C:\Program Files (x86)\Java\jdk{version}).

- **Spark**: You need to download Spark and copy it to one of the folders. The version you download should be compatible with Mobius. Hence, check the Mobius versioning policy to identify which version of Spark you can use. You have used Mobius Nuget v2.0.2, which matches Spark 2.0.2; hence, you have to download the same version of Spark from http://spark.apache.org/releases/spark-release-2-0-2.html. Once downloaded, I've extracted it to the C:\SparkMobius\spark-2.0.2-bin-hadoop2.7 folder. Now set this folder path as a new environment variable called SPARK_HOME.

- **Mobius**: Download the appropriate version (in our case 2.0.2) of Mobius from https://github.com/Microsoft/Mobius and put it in the C:\SparkMobius\Mobius-2.0.200 folder. I've downloaded a Mobius source. Navigate to the build folder inside the Mobius-2.0.200 folder and start build.cmd, which will build Mobius and get all the necessary utilities to run it in local mode. And finally, create a new environment variable, SPARKCLR_HOME, and point it to the build\runtime folder inside your Mobius folder; hence for my case it is C:\SparkMobius\Mobius-2.0.200\build\runtime.

- **Winutils.exe**: When running on Windows, Spark needs this utility. If you have built Mobius, then you already have it in the Mobius folder under tools\winutils. Create a new environment variable, HADOOP_HOME, and point it to the winutils.exe directory, which in our case is C:\SparkMobius\Mobius-2.0.200\build\tools\winutils.

Once all the dependencies are set, go to the command prompt and execute the following commands to verify that the environment variables are set correctly.

```
echo JAVA_HOME=%JAVA_HOME% & echo.SPARK_HOME=%SPARK_HOME% & echo.SPARKCLR_
HOME=%SPARKCLR_HOME% & echo.HADOOP_HOME=%HADOOP_HOME%
```

The output of this command on my machine looks like Figure 8-10.

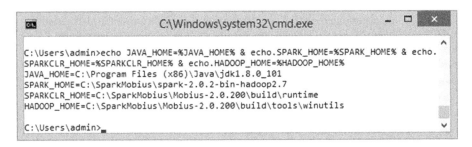

Figure 8-10. *Environment variables*

After all the environment variables are set correctly, go into the command prompt and type the following command.

```
%SPARKCLR_HOME%\scripts\sparkclr-submit.cmd --exe SparkClrWordCount.exe C:\
SparkClrWordCount\bin\debug file:///C:\data\the_adventures_of_sherlock_
holmes.txt
```

You are using the sparkclr-submit.cmd to execute the program in local mode. Also, note that you are passing the exe parameter, which is name of our exe, followed by path to the exe folder, and finally, a file from Project Gutenberg from which you want to run the word count program. If you wish, you can use any other text file instead. The result of the sparkclr-submit is shown in Figure 8-11 (note that this is not the complete output, but only a few lines from the beginning of the script and a few lines at the end).

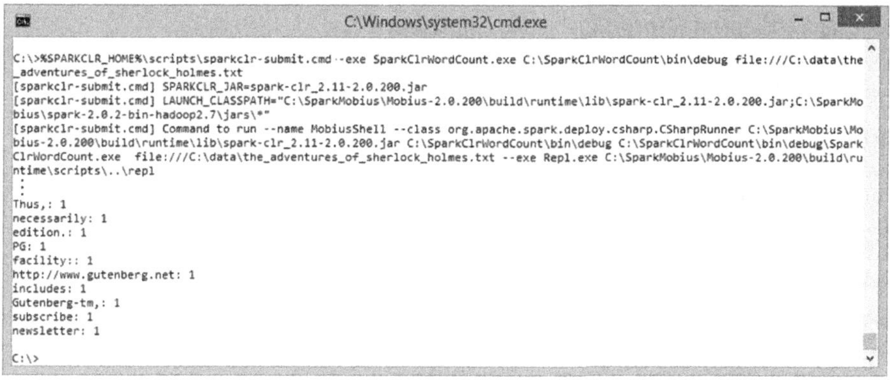

```
C:\Windows\system32\cmd.exe                                                    _ □ ×

C:\>%SPARKCLR_HOME%\scripts\sparkclr-submit.cmd --exe SparkClrWordCount.exe C:\SparkClrWordCount\bin\debug file:///C:\data\the
_adventures_of_sherlock_holmes.txt
[sparkclr-submit.cmd] SPARKCLR_JAR=spark-clr_2.11-2.0.200.jar
[sparkclr-submit.cmd] LAUNCH_CLASSPATH="C:\SparkMobius\Mobius-2.0.200\build\runtime\lib\spark-clr_2.11-2.0.200.jar;C:\SparkMo
bius\spark-2.0.2-bin-hadoop2.7\jars\*"
[sparkclr-submit.cmd] Command to run --name MobiusShell --class org.apache.spark.deploy.csharp.CSharpRunner C:\SparkMobius\Mo
bius-2.0.200\build\runtime\lib\spark-clr_2.11-2.0.200.jar C:\SparkClrWordCount\bin\debug C:\SparkClrWordCount\bin\debug\Spark
ClrWordCount.exe  file:///C:\data\the_adventures_of_sherlock_holmes.txt --exe Repl.exe C:\SparkMobius\Mobius-2.0.200\build\ru
ntime\scripts\..\repl
       :
       :
Thus,: 1
necessarily: 1
edition.: 1
PG: 1
facility:: 1
http://www.gutenberg.net: 1
includes: 1
Gutenberg-tm,: 1
subscribe: 1
newsletter: 1

C:\>
```

Figure 8-11. SparkClr-Shell result for word count

So, this is how you can execute your application in local mode. Also, it is possible to directly execute a Spark application from inside Visual Studio. It is called debug mode. Follow these instructions to run it.

1. Open a command prompt and navigate to the %SPARKCLR_HOME%\scripts folder.

2. Execute the sparkclr-submit.cmd debug command. This will start sparkclr-submit in debug mode at default port 5567.

3. If you want run it on a different port, then you can specify the same, but make sure that you add the following settings in your application config if you are not running on the default port.

    ```
    <appSettings>
    <add key="CSharpWorkerPath" value="/path/to/
    driverprogram/CSharpWorker.exe"/>
    <add key="CSharpBackendPortNumber"  value="port_number_
    from_previous_step"/>
    </appSettings>
    ```

4. Hit F5 from Visual Studio and you should be able to run the application. It generates the same output as in the console execution.

■ **Note**   While working with Mobius, if you find yourself in situation where sparkclr-submit does not work in debug mode and it gives exceptions when starting up, then you might have to set the execution memory option in the sparkclr-submit.cmd file's line 81 (in version 2.0.2) and add option "-Xmx512m" at the end of line.

Running in HDInsight Spark Cluster

Running a Mobius Spark application in an HDInsight Spark cluster is relatively easier than running in local mode. Also, you can use the same steps mentioned in this section to run on any Linux-based Spark cluster using YARN. The following are the steps to run a Mobius application on a YARN-based Spark cluster.

1. Mobius on a Linux cluster requires Mono 4.2 or higher to run C# code. First, check the version installed on your Linux node. Use PuTTY to get SSH to your cluster's master node. And then run the mono --version command to check the installed version. If it is not 4.2, then you might need to install/upgrade it. Use the sudo apt-get install mono-complete command to do so.

2. After you have Mono installed, you need to get Mobius release bits onto the cluster. Use the following command script to create a Mobius directory and get the latest bits matching your Spark version.

    ```
    >mkdir mobius
    >cd mobius
    >wget https://github.com/Microsoft/Mobius/releases/
    download/v2.0.200/spark-clr_2.11-2.0.200.zip
    >unzip spark-clr_2.11-2.0.200.zip
    >export SPARKCLR_HOME=/home/sshuser/mobius/runtime
    ```

3. Upload your application code. Create a new examples directory in the user home directory using the mkdir examples command in putty.

4. Execute the following command in the Windows command prompt to upload zip. This zip typically containing all the files from your bin/debug folder.

    ```
    C:\Program Files (x86)\PuTTY> pscp
    C:/SparkClrWordCount.zip sshuser@hdi-ssh.
    azurehdinsight.net:/home/sshuser/examples/
    ```

5. In the PuTTY folder, change the username (sshuser) and cluster URL (hdi-ssh.azurehdinsight.net) to match your cluster URL.

6. After uploading the zip, unzip it from PuTTY using the following command.

```
> unzip /home/sshuser/examples/SparkClrWordCount.zip
```

7. Once you have everything on the cluster, you can use sparkclr-submit.sh to submit a Spark job. Also, you need to give execute permission to the sparkclr-submit.sh and exe.

```
cd /home/sshuser/mobius/runtime/scripts
chmod +x sparkclr-submit.sh
chmod +x /home/sshuser/examples/SparkClrWordCount.exe
```

8. And finally, submit it using the following command.

```
./sparkclr-submit.sh --master yarn --deploy-mode
client --exe SparkClrWordCount.exe /home/sshuser/examples
```

Jupyter Notebook

The Jupyter Notebook is an open source web application in which you can write live code, and execute, save, and share it. It is an ideal tool for learning and quick testing. It is a feature-rich tool that can be used for live work as well. You can execute multiple notebooks and inherit one into another, creating complex notebooks that can do anything that you want with your Spark cluster. Jupyter Notebook allows code, equations, visualization, and explanatory text.

The HDInsight Spark cluster already has Jupyter Notebook installed on it. To open Jupyter Notebook, go to the cluster blade in HDInsight and click the cluster dashboards, which open another blade, as shown in Figure 8-12. From there, you can open Jupyter Notebook.

Figure 8-12. *HDInsight Spark cluster dashboards*

Another way is to directly navigate to https://{clustername}.azurehdinsight. net/jupyter to go to the Jupyter Notebook dashboard. No matter how you create a notebook, Spark context is created for you to use.

■ **Note**   Jupyter originally stood for Julia, Python, and R, which were the main languages used in the notebook.

On the Jupyter Notebook page, click New and select PySpark kernel, which is the default kernel. A *kernel* is a program that runs and interprets your code. The HDInsight Spark cluster provides four different kernels: PySpark, PySpark3, Spark, and SparkR, which you can see in Figure 8-13. PySpark and PySpark3 exposes the Spark programming model to Python. The Spark kernel uses Scala and SparkR uses the R language.

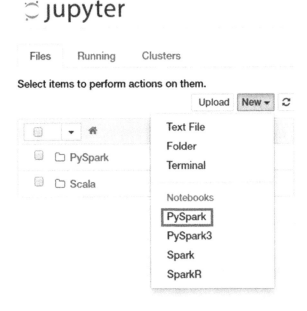

Figure 8-13. *Jupyter Notebook kernel*

Once you are on new notebook, you can enter code or mark down in it. Code will be executed as per the kernel you selected. To demonstrate notebook usage, let's create a Spark program to find the top ten words used in the Davinci text file (available on wasb storage at was:///example/data/gutenberg/davinci.txt).

9.   Open a new PySpark notebook.

10. Load a text file as RDD using the following snippet.

```
textLines = spark.sparkContext.textFile('wasb:///
example/data/gutenberg/davinci.txt')
```

11. Generate an RDD of words by splitting rows of text by space.

```
words = textLines.flatMap(lambda line: line.split(' '))
```

12. To count words, create a key-value pair of words and its total count.

```
wordPair = words.map(lambda word: (word, 1))
counts = wordPair.reduceByKey(lambda a, b: a + b)
```

13. Apply the takeOrdered action to find the top 10 words in descending order by word count.

```
counts.takeOrdered(10, lambda a: -a[1])
```

Figure 8-14 shows the complete notebook with the result of running it on an HDInsight Spark cluster.

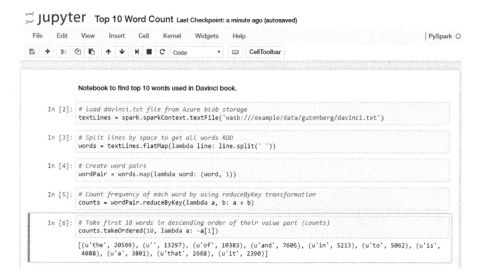

Figure 8-14. *Jupyter Notebook finds top ten words in Davinci text file*

Spark UI

Spark UI, or Web UI, is the web interface that you use to learn what is going on inside your Spark jobs. You can drill down into job tasks, executor details, DAG (direct acyclic graph) visualization, the input/output of each stage of the jobs, and so forth. This is very useful information when trying to find out where any bottlenecks are and in understanding how things work internally. Execution DAG shows the chain of RDD dependencies. Developers can quickly see whether an RDD operation is performing in the right manner or not.

To open the Spark UI, you should have a running job or a past job. To open a past job, go to the cluster dashboard and select Spark History Server. For a running job, open YARN from the cluster dashboard and select the running application. If you want to run a previous Jupyter Notebook, it keeps a running job until you close the notebook. If you followed the last section and have a notebook open with the kernel connected, then you should see a YARN similar to the one shown in Figure 8-15, where the name of the application is livy-session-9 and the status is RUNNING.

Show 20 ▾ entries

ID	User	Name	Application Type	Queue	Application Priority	StartTime	FinishTime	State
application_1488732652073_0014	livy	livy-session-9	SPARK	default	0	Mon Mar 6 01:10:01 +0550 2017	N/A	RUNNING
application_1488732652073_0013	livy	livy-session-8	SPARK	default	0	Mon Mar 6 00:55:11 +0550 2017	Mon Mar 6 00:56:44 +0550 2017	FINISHED
application_1488732652073_0012	livy	livy-session-7	SPARK	default	0	Mon Mar 6 00:29:50 +0550 2017	Mon Mar 6 00:54:43 +0550 2017	FINISHED

Figure 8-15. *YARN application execution history*

But this is not the Spark UI. To open the Spark UI, click the application ID, and from the Application Overview page, click the Tracking URL field, as shown in Figure 8-16.

User:	livy
Name:	livy-session-9
Application Type:	SPARK
Application Tags:	livy-session-9-xfryrfxu
Application Priority:	0 (Higher integer value indicates higher priority)
Elapsed:	40mins, 23sec
Tracking URL:	ApplicationMaster
Log Aggregation Status	NOT_START

Figure 8-16. *Application overview*

The Spark UI shows jobs launched by Jupyter Notebook. In my case, there is only one job launched in the current session, as shown in Figure 8-17.

Figure 8-17. Spark UI jobs view

There are several tabs available, such as Jobs, Stages, Storage, Environment, Executors, and SQL. Each tab gives different details about the job and its execution. The Jobs view shows the takeOrdered action, which you executed in last section, took 7 seconds to execute and there were two stages. To understand what these two stages are, click takeOrdered. Figure 8-18 shows the stages of the task. As you can see, out of 7 seconds, 6 seconds went into the reduce by key action, because to find the top ten words, you need to find the frequency of each word, which is what takes most of the time. The result of this stage is 657KB of data, which needs the shuffle as well. You can drill down more on each stage by clicking stage description. This brings more matrices for you to better understand what is going on inside the stage.

Stages for All Jobs

Completed Stages: 2

Completed Stages (2)

Stage Id	Description	Submitted	Duration	Tasks: Succeeded/Total	Input	Output	Shuffle Read	Shuffle Write
1	takeOrdered at <stdin>:2 +details	2017/03/05 19:40:40	0.5 s	2/2			657.2 KB	
0	reduceByKey at <stdin>:2 +details	2017/03/05 19:40:34	6 s	2/2	64.0 KB			657.2 KB

Figure 8-18. *Stages*

The Storage tab shows any RDD that is cached by memory. To demonstrate this behavior, open Jupyter Notebook and write the following Python code.

```
textLines = spark.sparkContext.textFile('wasb:///example/data/gutenberg/
davinci.txt').cache()
textLines.count()
```

Notice the cache() method call at the end of the first line. This call essentially tells Spark to keep the RDD in memory for further processing. This significantly improves performance. So, after executing the count action on the textLines RDD, if you try to find the top ten words, it executes quickly. Try executing the following line in another cell in the notebook, and then navigate to Spark UI.

```
textLines.flatMap(lambda line: line.split(' ')).map(lambda word: (word, 1)).
reduceByKey(lambda a, b: a + b).takeOrdered(10, lambda a: -a[1])
```

In my case, it previously took 7 seconds to finish the job; but after executing the count and then counting the top ten records, it is executed in less than a second, which you can see in Figure 8-19. This is how Spark can really improve iterative algorithms performance.

Job Id	Description	Submitted	Duration	Stages: Succeeded/Total	Tasks (for all stages): Succeeded/Total
2	takeOrdered at <stdin>:1	2017/03/05 20:45:05	0.9 s	2/2	4/4
1	count at <stdin>:2	2017/03/05 20:44:55	0.6 s	1/1	2/2
0	takeOrdered at <stdin>:2	2017/03/05 19:40:34	7 s	2/2	4/4

Figure 8-19. *Jobs view after cache*

Now if you go to Storage view, you should see one RDD cached.

The other tabs are Environment and Executors, which as their names suggest, give you information about a cluster's environment and executors.

DataFrames and Datasets

DataFrames is also like RDD, immutable distributed collection of data. But unlike an RDD, it is organized as named columns. You can think of DataFrames as a table in a relational database. It makes working with data even easier because you don't have to rely on an array index to identify a column when working with CSV or JSON data. It allows developers to impose structure onto a distributed collection of data. It provides higher-level abstraction and provides a domain-specific API to manipulate data. DataFrames is used easily with database tables, JSON, CSV, or serialized files. It is a higher-level API compared to RDD, which has benefits in terms of both storage and computation because Spark can decide in which format it needs to be handled. Under the hood, the Catalyst optimizer and Tungsten execution engine optimize applications in a way that is not possible with RDD, such as storing in raw binary form. The Tungsten execution engine chooses CPU and memory optimization over the network or I/O. It is designed to not waste a CPU cycle in SQL query execution and it works directly on the byte level. Catalyst optimizer is a query plan optimizer, which takes advantage of the Scala language feature, including pattern matching and runtime metaprogramming to allow developers to specify complex relational optimizations concisely.

A new Datasets API was introduced in Spark 1.6, making it even easier to work with data. The Datasets API allow type safety for structured data, and like DataFrames, it also takes advantage of Spark's Catalyst optimizer. Datasets also leverages Tungsten's fast in-memory encoding, and with compile-time type safety, which means an application can be checked for errors before it is deployed to a cluster. Another benefit of the Dataset API is the reduction in memory of object. As Spark understands the layout of objects in Datasets, it can create a more optimal layout in memory when caching Datasets. Encoders are highly optimized and use runtime code generation to build custom bytecode for serialization and deserialization. It performs significantly faster than Java or Kyro. Like RDD, DataFrames and Datasets also have their own APIs to make it easy to work with them.

Let's move to next section, where you will use DataFrames and Datasets.

The following is a sample DataFrame-based Mobius application to read data from an MS SQL database and find the total number of rows by using the DataFrame API.

```
static void Main(string[] args)
{
    LoggerServiceFactory.SetLoggerService(Log4NetLoggerService.Instance);
//this is optional - DefaultLoggerService will be used if not set
    var logger = LoggerServiceFactory.GetLogger(typeof(JdbcDataFrameExample));

    var connectionString = args[0];
    var tableName = args[1];

    var sparkConf = new SparkConf();

// Create Spark Context
    var sparkContext = new SparkContext(sparkConf);

// Create SQL Context
    var sqlContext = new SqlContext(sparkContext);

// Create DataFrame fom JDBC connection
    var df = sqlContext
                .Read()
                .Jdbc(connectionString, tableName,
        new Dictionary<string, string> {
                        { "driver", "com.microsoft.sqlserver.jdbc.
                        SQLServerDriver" }});

    // Perform row count
    var rowCount = df.Count();
    logger.LogInfo("Row count is " + rowCount);
    sparkContext.Stop();
}
```

To run the preceding code in local mode, use the following sparkclr-submit. Note that you are using a local instance of MS SQL Server, which has a Temp database and a MyTable table.

```
%SPAKRCLR_HOME%\scripts\sparkclr-submit.cmd --exe SparkJdbc.exe C:\
SparkJdbc\bin\debug "jdbc:sqlserver://localhost;databaseName=Temp;user=MyUse
rName;password=myPassword;" "MyTable"
```

The results are the number of rows in the MyTable table with a bunch of log items.

Spark SQL

So far, you have seen how to work with RDD and how to use different transformations and actions to get results from data. But sometimes it is easy to express the operations in terms of SQL rather than providing transformations and actions. Also, not all developers are comfortable working with transformation and actions. Spark SQL provides easy-to-program abstraction using SQL over data with Spark. Compared to HiveQL, this is standard SQL. Spark SQL supports the SQL 2003 standard. Internally, everything is broken down to Map and Reduce jobs.

the first example reads a local JSON file and then filters it using a SQL query rather than transformations.

Assume that you have the following data in a JSON text file, with each line containing a single record.

```
{"name":"Michael" }
{"name":"Judas", "age":35}
{"name":"Andy", "age":30}
{"name":"Justin","age": 19}
{"name":"Jordan", "age":60}
```

You need to find all the records where age is greater than 20 and the name starts with letter J. To work with Spark SQL, you need to create SQLContext along with SparkContext. The following is code to read a JSON file and then do the filtering using SQL query.

```
static void Main(string[] args)
{
    LoggerServiceFactory.SetLoggerService(Log4NetLoggerService.Instance);
    var logger = LoggerServiceFactory.GetLogger(typeof(Program));

    var sparkConf = new SparkConf();
    sparkConf.SetAppName("myapp");
    var sparkContext = new SparkContext(sparkConf);
    var sqlContext = new SqlContext(sparkContext);

    try
    {
        logger.LogDebug(args[0]);

        // Read file
        var df = sqlContext.Read().Json(args[0]);
        df.Show();

        // Create temporary table
        df.RegisterTempTable("TempPersonTable");
```

```
        // Execute SQL query
        var filteredDf = sqlContext
                    .Sql("SELECT * FROM TempPersonTable where age > 20 AND
                    name like 'J%'");

filteredDf.Show();
    }
    catch (Exception ex)
    {
        logger.LogException(ex);
    }
    sparkContext.Stop();
}
```

Submit this to local Spark using the following command.

```
%SPAKRCLR_HOME%\scripts\sparkclr-submit.cmd --exe SparkSQLApp.exe
C:\SparkSQL\b
in\Debug file:///C:/data.json
```

It generates the following output.

```
+---+------+
|age|  name|
+---+------+
| 35| Judas|
| 60|Jordan|
+---+------+
```

In a similar fashion, you can query any DataFrames using Spark SQL. Note that even if you know transformations and actions, Spark SQL may give better performance due to that internal optimization applied on DataFrames.

Summary

In this chapter, you explored concepts related to Apache Spark. Apache Spark is an all-in-one technology to do batch processing, stream analytics, machine learning, or graph data processing. It provides very easy-to-use developer APIs that allow developers to write code that can run in parallel on a cluster. It gives dual benefits in terms of better performance over MapReduce and is easier to code than other Hadoop technologies. SQL and R users can also use Spark for data processing.

In-memory distributed collection of data makes it the best suitable for iterative algorithms and fast interactive queries. The Mobius project has an almost identical to JAVA API used in C# and F#, which opens Spark to the whole .NET community. Mobius can be used on Windows and Linux clusters, easily giving power to a .NET developer to build Spark applications for any Spark cluster. Spark SQL provides rich SQL-based API to process data, which can become cumbersome when using transformations and actions. Overall, Spark is a complete package for today's real-world big data processing application.

Index

Get the eBook for only $5!

Why limit yourself?

With most of our titles available in both PDF and ePUB format, you can access your content wherever and however you wish—on your PC, phone, tablet, or reader.

Since you've purchased this print book, we are happy to offer you the eBook for just $5.

To learn more, go to http://www.apress.com/companion or contact support@apress.com.

Apress®

Printed in the United States
By Bookmasters